Progress in Probability and Statistics
Volume 13

Series Editor
Murray Rosenblatt

Seminar on Stochastic Processes, 1986

E. Çınlar,
K.L. Chung,
R.K. Getoor,
Editors

J. Glover,
Managing Editor

1987

Birkhäuser
Boston · Basel · Stuttgart

E. Çınlar
Civil Engineering Department
Princeton University
Princeton, NJ 08544
U.S.A.

K.L. Chung
Department of Mathematics
Stanford University
Stanford, CA 94305
U.S.A.

R.K. Getoor
Department of Mathematics
University of California
La Jolla, CA 92093
U.S.A.

J. Glover (Managing Editor)
Department of Mathematics
University of Florida
Gainesville, FL 32611
U.S.A.

ISSN: 0892-063X

CIP-Kurztitelaufnahme der Deutschen Bibliothek
Seminar on Stochastic Processes:
Seminar on Stochastic Processes . . .—Boston ; Basel ;
Stuttgart : Birkhäuser
6. 1986 (1987).
(Progress in probability and statistics ; Vol. 13)
ISBN-13: 978-1-4684-6753-6 e-ISBN-13: 978-1-4684-6751-2
DOI: 10.1007/978-1-4684-6751-2
NE: GT

Softcover reprint of the hardcover 1st edition 1987

ISBN-13: 978-1-4684-6753-6

9 8 7 6 5 4 3 2 1

FOREWORD

The 1986 Seminar on Stochastic Processes was held at the University of Virginia, Charlottesville, in March. It was the sixth seminar in a continuing series of meetings which provide opportunities for researchers to discuss current work in stochastic processes in an informal atmosphere. Previous seminars were held at Northwestern University, Evanston and the University of Florida, Gainesville.

The participants' enthusiasm and interest have resulted in stimulating and successful seminars. We thank them for it, and we also thank those participants who have permitted us to publish their research here.

The seminar was made possible through the generous support of the Office of Naval Research (Contract # A86-4633-P) and the University of Virginia. We are grateful for their support. The participants were welcomed to Virginia by S. J. Taylor, whose store of energy and organizing talent resulted in a wonderful reunion of researchers. We extend to him our warmest appreciation for his efforts; his hospitality makes us hope that we can someday return to Virginia for another conference.

J. G.

Gainesville, 1986

TABLE OF CONTENTS

Green's Function for a Ball

K. L. Chung*

Let $B = B(a,r)$ be the open ball with center a and radius r in R^d, $d \geq 3$; ∂B its boundary sphere. For $x \neq a$, its inversion with respect to B is defined to be

$$x^* = a + \frac{r^2}{|x - a|^2}(x - a). \tag{1}$$

We have

$$|x - a||x^* - a| = r^2; \tag{2}$$

from which it follows that the mapping is involutary: $(x^*)^* = x$. Also we have from (1):

$$\begin{aligned} |x^* - y|^2 &= |a - y|^2 + \frac{2r^2}{|x - a|^2}(a - y, x - a) \\ &\quad + \frac{r^4}{|x - a|^4}|x - a|^2; \end{aligned} \tag{3}$$

$$\begin{aligned} |x^* - y|^2|x - a|^2 &= |x - a|^2|y - a|^2 \\ &\quad - 2r^2(x - a, y - a) + r^4. \end{aligned}$$

* Research supported by AFOSR Grant 85-0330.

The right-hand member of (3) being symmetric in (x,y), we see that

(4) $$|x^* - y|^2|x - a|^2 = |y^* - x|^2|y - a|^2.$$

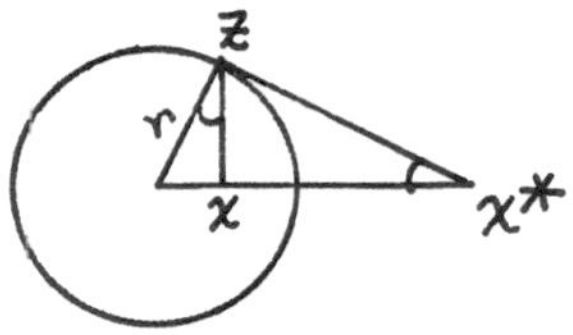

Figure

It follows from (2) and the Figure that if $z \in \partial B$, we have by similar triangles:

(5) $$\frac{|x - a|}{r} = \frac{r}{|x^* - a|} = \frac{|x - z|}{|x^* - z|};$$

namely for any $x \in B$, $x \neq a$:

(6) $$\partial B = \{z \in R^d : \left|\frac{z - x}{z - x^*}\right| = \frac{|x - a|}{r}\}.$$

It is clear that we may put $a = 0$ by the mapping $x \rightsquigarrow x - a$. Next, we put

(7) $$f(x,y) = |x||x^* - y| = |y||y^* - x|,$$

and compute the key formula:

(8) $$f(x,y)^2 = |x|^2 \left|\frac{r^2 x}{|x|^2} - y\right|^2 = r^4 - 2r^2(x,y) + |x|^2|y|^2$$
$$= r^2|x - y|^2 + (r^2 - |x|^2)(r^2 - |y|^2).$$

We now introduce

$$(9) \qquad U(x,y) = \frac{A_d}{|x - y|^{d-2}}, \quad (x,y) \in R^d \times R^d$$

with $U(x,x) = +\infty$, where

$$(10) \qquad A_d = \frac{\Gamma(\frac{d}{2} - 1)}{4\pi^{d/2}}.$$

The function u is known as the Green's function for R^d. The Green's function for B is the function G defined on $B \cup (\partial B)$ as follows:

$$(11) \quad G(x,y) = A_d\{(\frac{1}{|x - y|})^{d-2} - (\frac{r}{|x||x^* - y|})^{d-2}\}.$$

Since $|x||x^* - y| > r|x - y|$ by (8), it follows that

$$(12) \qquad 0 < G(x,y) < U(x,y)$$

in $B \times B$; while

$$(13) \qquad G(x,z) = 0$$

on $B \times \partial B$ by (5). For each $y \in B$, it can be verified that $U(\cdot,y) - G(\cdot,y)$ is harmonic in $B - \{y\}$ and takes on the boundary value of $U(\cdot,y)$ on ∂B. The last two properties uniquely determine G, and is its raison d'être in classic potential theory. The constant A_d has its significance, but since it plays no role in what follows it is sometimes omitted in the difference of U.

The role of the radius r is not so clear. However, a

straight forward computation shows that if we denote temporarily the G in (11) by G_r, we have the reduction formula:

$$G_r(x,y) = \frac{1}{r^{d-2}} G_1\left(\frac{x}{r},\frac{y}{r}\right). \tag{14}$$

This permits us to concentrate on B = B(0,1) and the G in (11) with r = 1. It follows from (4) that G is symmetric in (x,y):

$$G(x,y) = G(y,x).$$

We shall denote the distance of x in B to ∂B by $\delta(x) = 1 - |x|$.

Proposition 1. We have

$$\tfrac{1}{4}\min\left(\frac{1}{|x-y|^{d-2}}, \frac{\delta(x)\delta(y)}{|x-y|^{d}}\right) < \frac{G(x,y)}{A_d} \tag{15}$$

$$< \min\left(\frac{1}{|x-y|^{d-2}}, \frac{4(d-2)\delta(x)\delta(y)}{|x-y|^{d}}\right)$$

Proof. The inequality on the right with the first term under min is just (12). Now we write

$$\frac{G(x,y)}{A_d} = \frac{f(x,y)^{d-2} - |x-y|^{d-2}}{|x-y|^{d-2} f(x,y)^{d-2}}. \tag{16}$$

Since $f(x,y) > |x - y|$ by (8) with r = 1, the numerator in (16) is less than

$$(d - 2)(f(x,y) - |x - y|)f(x,y)^{d-3}$$

$$< (d - 2)(f(x,y)^2 - |x - y|^2)f(x,y)^{d-4}$$

$$< 4(d - 2)\delta(x)\delta(y)f(x,y)^{d-4}$$

Since $1 - |x|^2 < 2\delta(x)$. Substituting into (16) and using $f(x,y) > |x - y|$ again, we obtain the inequality on the right of (15) with the second term under the min.

On the other hand, the numerator in (16) is greater than

$$(f(x,y) - |x - y|)f(x,y)^{d-3}$$

$$> (f(x,y)^2 - |x - y|^2)\ \tfrac{1}{2}\ f(x,y)^{d-4}.$$

Substituting into (16) we obtain

$$\frac{G(x,y)}{A_d} > \frac{f(x,y)^2 - |x - y|^2}{2|x - y|^{d-2}f(x,y)^2}.$$

Since for $A > 0$, $B > 0$, we have $\frac{A}{A + B} > \tfrac{1}{2}\min(1,\frac{A}{B})$, it follows from (8) that

$$\frac{G(x,y)}{A_d} > \frac{1}{4|x - y|^{d-2}}\min\left(1,\frac{(1 - |x|^2)(1 - |y|^2)}{|x - y|^2}\right)$$

Since $1 - |x|^2 > \delta(x)$, this implies the left-hand inequality in (15). ∎

Proposition 1 can be blown up as follows. Put

(17) $F(x,y)=\min\{\frac{1}{|x - y|^{d-2}}, \frac{\delta(x)}{|x - y|^{d-1}}, \frac{\delta(y)}{|x - y|^{d-1}}, \frac{\delta(x)\delta(y)}{|x - y|^{d}}\}$

Proposition 2. There exist constants A_1 and A_2 depending only on d such that for all $(x,y) \in B \times B$:

(18) $$A_1F(x,y) < G(x,y) < A_2F(x,y)$$

Proof. From here on we shall use A to denote any changeable constant depending only on d. Let us first show that if

(19) $$G(x,y) < A \min \left(\frac{1}{|x - y|^{d-2}}, \frac{\delta(x)\delta(y)}{|x - y|^{d}}\right)$$

then

(20) $$G(x,y) < 4A \frac{\delta(x)}{\delta(y)|x - y|^{d-2}}$$

This is trivial if $2\delta(x) > \delta(y)$. If $\delta(y) > 2\delta(x)$, then

$$|x - y| > \delta(y) - \delta(x) > \tfrac{1}{2}\,\delta(y)$$

Hence

$$G(x,y) < A\frac{\delta(x)\delta(y)}{|x - y|^{d}} \cdot \frac{4|x - y|^2}{\delta(y)^2} = 4A\frac{\delta(x)}{\delta(y)|x - y|^{d-2}}$$

Thus (20) is true. It follows from this and (19) that

$$G(x,y)^2 < A\frac{\delta(x)}{\delta(y)|x - y|^{d-2}}\,\frac{\delta(x)\delta(y)}{|x - y|^{d}}$$

or

(21) $$G(x,y) < A\frac{\delta(x)}{|x - y|^{d-1}}.$$

This is then also true when $\delta(x)$ is replaced by $\delta(y)$, by the symmetry of G. Hence the right-hand inequality of (18) is true. If we wish we can also insert the right-hand member of (20), and another term obtained from it interchanging x and y, inder the min in (17) for the definition of F. The left-hand inequality then follows automatically from the left-hand inequality in (15).

We now make the important observation that Proposition 2 is invariant when B(0,1) is replaced by B(0,r), provided of course that $\delta(x)$ is interpreted as the distance from x to $\partial B(0,r)$. For if we write this distance as $\delta_r(x)$, then $\delta_r(x) = r - |x| = r\delta_1(\frac{x}{r})$, so that

(22) $$F(\frac{x}{r},\frac{y}{r}) = r^{d-2}F(x,y).$$

Therefore by (14), the inequalities in (18) are unchanged when B(0,1) is replaced by B(0,r). Similarly, the constant A in the next proposition does not depend on r. The next result originated with Brossard.

Proposition 3. There exists a constant A depending only on d such that

(23) $$\frac{G(x,y)G(y,z)}{G(x,z)} < A\frac{U(x,y)U(y,z)}{U(x,z)}$$

for all x, y and z in B.

Proof. We have by (18) and (20):

$$G(x,y)G(y,z) < A\frac{\delta(x)\delta(z)}{|x - y|^d|y - z|^{d-2}};$$

hence also by symmetry

$$G(x,y)G(y,z) < A\frac{1}{|x-y|^{d-2}|y-z|^{d-2}}\min\{1,\frac{\delta(x)\delta(z)}{|x-y|^2},\frac{\delta(x)\delta(z)}{|y-z|^2}\}.$$

On the other hand we have by (15):

$$G(x,z) > \frac{A}{|x - z|^{d-2}}\min\{1,\frac{\delta(x)\delta(z)}{|x - z|^2}\}.$$

Hence if $\delta(x)\delta(z) > |x - z|^2$, the left member of (23) does not exceed

$$\frac{1}{A}\left|\frac{x - z}{(x - y)(y - z)}\right|^{d-2};$$

if $\delta(x)\delta(z) < |x - z|^2$, it does not exceed

$$A\left|\frac{x - z}{(x - y)(y - z)}\right|^{d-2}\min\left\{\left|\frac{x - z}{x - y}\right|^2, \left|\frac{x - z}{y - z}\right|^2\right\}.$$

Since $|x - y| + |y - z| > |x - z|$, the last-written min does not exceed 4. This establishes (23).

Let $w \in \partial B$, then

$$\lim_{z\to w}\frac{G(x,z)}{\delta(z)} = -\frac{\partial}{\partial n_w}G(x,w) = K(x,w)$$

where $\frac{\partial}{\partial n_w}$ denotes the outward normal derivative at w, since $G(x,w) = 0$ for $x \in B$, by (13). The function $K(\cdot,\cdot)$ defined on $B \times \partial B$ is known as Poisson's kernel. Dividing the left member of (23) by $\delta(z)$ in both numerator and denominator, and letting $z \to w$, we obtain

$$(24) \qquad G^w(x,y) \overset{\Delta}{=} \frac{G(x,y)K(y,w)}{K(x,w)} \leq C\frac{U(x,y)U(y,w)}{U(x,w)}.$$

However, for the ball $B(0,r)$, Poisson's kernel is known explicitly:

$$(25) \qquad K(x,z) = \frac{A_d d}{r} \frac{r^2 - |x|^2}{|x - z|^d}$$

where A_d is given by (10). Hence (24) is trivial. From (24) we derive easily the inequality

$$(26) \qquad G^w(x,y) \leq A \max \left(\frac{1}{|x - y|^{d-2}}, \frac{1}{|y - w|^{d-2}}\right)$$

which is a fundamental estimate, also given by Brossard. His proof is quite different.

We now consider $B = B(0,r)$ in R^2. In this case the Green's function for B is given by

$$(27) \qquad G(x,y) = \frac{1}{\pi} \log \frac{|x||x^* - y|}{r|x - y|}, \quad (x,y) \in B \times B.$$

Then $G(x,y) > 0$, and $= 0$ if $x \in \partial B$ or $y \in \partial B$, as before. We put

$$(28) \qquad U_r(x,y) = \log \frac{3r}{|x - y|}$$

so that $U_r > \log \frac{3}{2} > \frac{2}{5}$ in $\overline{B} \times \overline{B}$. Put also

$$(29) \qquad \phi(x,y) = \frac{|x||x^* - y|}{r}.$$

Then by (8),

(30) $$\phi(x,y)^2 = |x - y|^2 + \frac{1}{r^2}(r^2 - |x|^2)(r^2 - |y|^2).$$

Hence $\phi^2 < (2r)^2 + r^2 = 5r^2$. Now we represent G as follows

(31) $$G(x,y) = \frac{1}{\pi}\{U_r(x,y) + \log \frac{\phi(x,y)}{3r}\}$$

The second term in the right member above is negative because $\sqrt{5}/3 < 1$, whereas the first term is bounded away from zero. This explains our choice of 3r in (28) rather than the usual one. An immediate consequence is that

(32) $$G(x,y) < \frac{1}{\pi}U_r(x,y).$$

Next, we have from (30)

(33) $$\log \phi(x,y) = \log |x - y| + \tfrac{1}{2} \log (1 + \psi(x,y))$$

where

(34) $$\psi(x,y) = \frac{(r^2 - |x|^2)(r^2 - |y|^2)}{r^2|x - y|^2}.$$

Since $r^2 - |x|^2 < 2r\delta(x)$, we have

(35) $$\psi(x,y) < \frac{4\delta(x)\delta(y)}{|x - y|^2}.$$

Since $\psi > 0$, $\log(1 + \psi) < \psi$; it follows from (33) and (35) that

(36) $$\log \phi(x,y) < \log |x - y| + \frac{2\delta(x)\delta(y)}{|x - y|^2}.$$

Observing that

(37) $$G(x,y) = \frac{1}{\pi} \log \frac{\phi(x,y)}{|x - y|}$$

we obtain from (36) that

(38) $$G(x,y) < \frac{2}{\pi} \frac{\delta(x)\delta(y)}{|x - y|^2}.$$

Continuing (32) and (38), we have (using $a \wedge b$ to denote min (a,b)):

(39) $$G(x,y) < \frac{1}{\pi}\{U_r(x,y) \wedge \frac{2\delta(x)\delta(y)}{|x - y|^2}\}.$$

Since $U_r > \frac{2}{5}$, this leads to the next proposition.

Proposition 4. In R^2, the Green's function. G for B(0,r) satisfies the following inequality:

(40) $$G(x,y) < \frac{1}{\pi} \log \frac{3r}{|x - y|}\{1 \wedge \frac{5\delta(x)\delta(y)}{|x - y|^2}\}$$

In contrast to Proposition 1 in the case $d > 3$, the inequality (40) cannot be reversed by changing the constants involved. In other words, there does not exist any constant $A > 0$ such that

(41) $$G(x,y) > A \log \frac{3r}{|x - y|}\{1 \wedge \frac{\delta(x)\delta(y)}{|x - y|^2}\}.$$

To see this let $0 < \varepsilon < 1$ and $\delta(x) = \delta(y) = |x - y| = \varepsilon r$. By (38), we have $G(x,y) < \frac{2}{\pi}$, whereas the right member of (41) reduces to $A \log \frac{3}{\varepsilon}$. It is not clear whether there exists a "sharp" estimate for G as in the case $d > 3$ above.

We proceed to an analogue for (26). For $d = 2$, the analogue of (25) is given by

$$(42) \qquad K(x,w) = \frac{1}{2\pi r} \frac{r^2 - |x|^2}{|x - w|^2}, \quad (x,w) \in B \times \partial B;$$

while $G^w(x,y)$ is defined as in (24). We have then

$$(43) \qquad G^w(x,y) < G(x,y)\frac{2\delta(y)}{\delta(x)} \frac{|x - w|^2}{|y - w|^2}$$

because $r^2 - |y|^2 < 2r\delta(y)$, $r^2 - |x|^2 > r\delta(x)$. Observe that

$$(44) \qquad 1 \wedge \frac{5\delta(x)\delta(y)}{|x - y|^2} < \frac{7\delta(x)}{\delta(y)}$$

This is trivial if $\delta(y) < 7\delta(x)$; otherwise it follows from $|x - y| > \delta(y) - \delta(x) > \frac{6}{7}\delta(y)$ and $5(\frac{7}{6})^2 < 7$. Using (44) in (40), we obtain

$$(45) \qquad G(x,y) < \frac{7}{\pi} \log\frac{3r}{|x - y|}\{\frac{\delta(x)}{\delta(y)} \wedge \frac{\delta(x)\delta(y)}{|x - y|^2}\}.$$

Therefore we have by (43)

$$(46) \quad G^w(x,y) < \frac{14}{\pi}\log\frac{3r}{|x - y|}\{1 \wedge \frac{\delta(y)^2}{|x - y|^2}\}\frac{|x - w|^2}{|y - w|^2}.$$

$$< \frac{14}{\pi}\log\frac{3r}{|x - y|}\{\frac{|x - w|^2}{|y - w|^2} \wedge \frac{|x - w|^2}{|x - y|^2}\}$$

because $\delta(y) < |y - w|$. The quantity between the braces above does not exceed 4, as shown in the proof of Proposition 3.

Proposition 5. For any $w \in \partial B$, we have

$$(47) \qquad G^w(x,y) < \frac{56}{\pi}\log\frac{3r}{|x - y|}.$$

In contrast to (26), this estimate of G^w does not depend on w.

Postscript. Some of the results above are implicit in the work by Z. Zhao, but the arrangements as well as formulations may be new. For instance, experts we consulted were not aware of the sharp form given in Proposition 2. It has since been proved for a bounded $C^{1,1}$ domain in R^d, $d > 3$, by Zhao (to appear in a book by us). According to some experts, once the results are established for a ball, geometrical transformations yield easily their extensions to a "reasonably smooth" domain. Although I am not privy to such arguments, this consideration makes it worthwhile to examine the case of a ball in detail.

K. L. Chung
Stanford University
Department of Mathematics
Stanford, CA 94305

ON THE IDENTIFICATION OF MARKOV

PROCESSES BY THE DISTRIBUTION OF HITTING TIMES

by

P.J. FITZSIMMONS

J. Glover [6,7] has recently provided a remarkable generalization of the celebrated Blumenthal, Getoor, McKean theorem [2] concerning the identification of Markov processes up to a time change. To state Glover's theorem let $X = (X_t, P^x)$ and $Y = (Y_t, Q^x)$ be right Markov processes on a common state space $(E, \mathcal{E})$. Let $\Delta \in E$ be a cemetary point used to render the resolvents of X and Y Markovian. Recall that Δ is a trap for X and for Y; the *lifetime* of X (resp. Y) is then $\zeta = \inf\{t: X_t = \Delta\}$(resp. $\eta = \inf\{t: Y_t = \Delta\}$). For $B \in \mathcal{E}$, let $T(B) = \inf\{t>0: X_t \in B\}$, $S(B) = \inf\{t>0: Y_t \in B\}$. Recall that X, for example, is *transient* provided its potential kernel U is proper.

(1) THEOREM. *Let* X *and* Y *be transient right processes. Suppose that* X *and* Y *have identical hitting probabilities; that is*

$$(2) \qquad P^x(T(B)<\infty) = Q^x(S(B)<\infty), \quad \forall x \in E, \forall B \in \mathcal{E}.$$

Then there exists a continuous additive functional (CAF) *of* X, *say* (A_t) , *which is strictly increasing and finite on* $[0,\zeta[$ *such that if*

$\tau(t) = \inf\{s: A_s > t\}$ *then* $(X_{\tau(t)}, P^x)$ *and* (Y_t, Q^x) *are equivalent processes.*

Recall that two processes are equivalent if they have the same finite dimensional distributions.

Actually Glover proved Theorem (1) in [7] under an additional hypothesis of absolute continuity. This restriction was lifted by the author in [3].

The condition (2) is a considerable weakening of the hypotheses of the Blumenthal, Getoor, McKean theorem which require that X and Y have identical hitting distributions; i.e. that $P^x(X_{T(B)} \varepsilon dz) = Q^x(Y_{S(B)} \varepsilon dz)$ for all $x \varepsilon E$ and all $B \varepsilon \mathcal{E} \vee \{\Delta\}$.

Our purpose in this note is to point out the amusing fact that if (2) is required for the α-subprocesses of X and Y (rather than for X and Y) for one fixed $\alpha > 0$, then X and Y must be equivalent processes.

(3) THEOREM. *Let* $X = (X_t, P^x)$ *and* $Y = (Y_t, Q^x)$ *be right Markov processes on a common state space* $(E, \mathcal{E})$. *Suppose that* X *and* Y *are both conservative in that* $P^x(X_t \varepsilon E) = Q^x(Y_t \varepsilon E) = 1$ *for all* $t \geq 0$. *Suppose that*

(4) $$P^x(\exp(-\alpha T(B))) = Q^x(\exp(-\alpha S(B))), \qquad \forall x \varepsilon E,\ \forall B \varepsilon \mathcal{E},$$

for one *fixed* $\alpha > 0$. *Then* X *and* Y *are equivalent processes.*

(5) REMARKS. a) In view of Glover's theorem (1) and Dynkin's characterization of α-excessive functions [1,II(5.1)], it is clear that (4) is equivalent to the statement

(6) X and Y have identical cones of α-excessive functions.

Of course, (2) is equivalent to the statement that X and Y have identical cones of (0-)excessive functions.

b) The conservation hypothesis on X and Y in (3) can be replaced by a more general condition. For suppose that the resolvents of X and Y are only subMarkovian so that the respective lifetimes ζ and η may be finite with positive probability. Regarded as processes with state space E_Δ, however, X and Y are conservative. In order that (4) hold for all $B \varepsilon \mathcal{E} \vee \{\Delta\}$ it is necessary and sufficient that (4) hold as written and that $P^x(\exp(-\alpha\zeta)) = Q^x(\exp(-\alpha\eta))$ for all $x \varepsilon E$. In this case X and Y are equivalent processes.

PROOF OF (3). It is assumed that X and Y are right processes in the sense of Getoor [4]. Let (U^β) and (V^β) denote the resolvents of X and Y respectively. By hypothesis $\alpha U^\alpha 1 \equiv 1 \equiv \alpha V^\alpha 1$. Let X^α (resp. Y^α) denote the α-subprocess of X (resp. Y). For example, we can obtain X^α by sending X to the cemetary state Δ at a random time which is independent of X and which follows the exponential distribution with paramenter α. Clearly (4) means that X^α and Y^α have identical hitting probabilities. Since X^α and Y^α are transient, Glover's theorem (1) applies; because of the simple relationship between X and X^α (resp. Y and Y^α), it follows that there is a CAF (A_t) of X such that if $\tau(t) = \inf\{s: A_s > t\}$ then $X^\alpha_{\tau(t)}$ and Y^α_t are equivalent processes. By an easy calculation this means that for all $x \varepsilon E$ and all bounded measurable f on E,

(7) $$V^\alpha f(x) = U^\alpha_A f(x),$$

where $U_A^\alpha f(x) = P^x \int_0^\infty e^{-\alpha t} f(X_t)\,dA_t$. Taking f = 1 in (7) we have

$$U_A^\alpha 1(x) = V^\alpha 1(x) = 1/\alpha, \quad \forall x \varepsilon E.$$

That is, (A_t) has the same finite α-potential as the CAF $B_t = t$. From [1,IV(2.13)] it now follows that $A_t = t, \forall t \geq 0$, a.s. P^x for all $x \varepsilon E$. Thus $\tau(t) \equiv t$ and so X^α and Y^α are equivalent processes. It is now immediate that X and Y are equivalent and so Theorem 3 is proved. □

There is a "dual" to (3) that follows easily from a result of Getoor and Glover [5] (which result is the dual of (1)). Let X and Y be conservative Borel right processes on $(E,\mathcal{E})$ with identical cones of α-excessive measures. In other words, the transient α-subprocesses X^α and Y^α have identical cones of excessive measures. By [5,(0.2)] there is a set $K \varepsilon \mathcal{E}$, polar for both X^α and Y^α (so also polar for both X and Y), and a positive Borel function c on E with $0 < c < \infty$ on $E\backslash K$, such that $U^\alpha f(x) = c(x)V^\alpha f(x)$ for $x\varepsilon E\backslash K$. Taking $f = 1$ and using $\alpha U^\alpha 1 = 1 = \alpha V^\alpha 1$ we see that $c = 1$ on $E\backslash K$. It follows easily that X restricted to $E\backslash K$ is equivalent to Y restricted to $E\backslash K$.

References

1. R.M. BLUMENTHAL and R.K. GETOOR. *Markov Processes and Potential Theory*. Academic Press, New York, 1968.
2. R.M. BLUMENTHAL, R.K. GETOOR, H.P. McKEAN, JR. Markov processes with identical hitting distributions. *Ill. J. Math.* *6* (1962), 402-420.

3. P.J. FITZSIMMONS. Markov processes with identical hitting probabilities. To appear in *Math. Zeit.*

4. R.K. GETOOR. *Markov Processes: Ray Processes and Right Processes.* Lecture Notes in Math. *440.* Springer-Verlag, Berlin-Heidelberg-New York, 1975.

5. R.K. GETOOR and J. GLOVER. Markov processes with identical excessive measures. *Math. Zeit. 184* (1983), 287-300.

6. J. GLOVER. Markov processes with identical hitting probabilities. *Trans. Am. Math. Soc. 275* (1983), 131-141.

7. J. GLOVER. Identifying Markov processes up to time change. *Seminar on Stochastic Processes 1982,* 171-194. Birkhäuser, Boston, 1983.

P.J. FITZSIMMONS
Department of Mathematical Sciences
The University of Akron
Akron, Ohio 44325

ON TWO RESULTS IN THE POTENTIAL

THEORY OF EXCESSIVE MEASURES

by

P.J. FITZSIMMONS

0. Introduction

Let (P_t) be the semigroup of a right Markov process and let m be an excessive measure for (P_t) (i.e., m is σ-finite and $mP_t \leq m$ for $t > 0$). As is well known, m can be uniquely decomposed as $m = m_i + m_p$ where m_i is *invariant* ($m_i P_t = m_i$ for $t > 0$), and m_p is *purely excessive* ($m_p P_t(f) \downarrow 0$ as $t \uparrow \infty$ for $f \geq 0$ with $m(f) < \infty$). The component m_p can be decomposed further:

$$(1) \qquad m_p = \int_0^\infty \nu_t \, dt,$$

where $(\nu_t\colon t > 0)$ is a family of σ-finite measures satisfying $\nu_t P_s = \nu_{t+s}$ for $s,t > 0$. The decomposition (1) seems to be well known (cf. [2]; see also [7] for a related result). A probabilistic proof of (1) is given in [3] by means of the stationary process associated with (P_t) and m. In [6], Getoor and Glover use (1) as an important step in their construction of the aforementioned stationary process. Actually, Getoor and Glover consider the more general (and more difficult) time inhomogeneous case, but even in the time homogeneous case their proof of (1) is involved.

Our purpose in this note is to give a simple direct proof of (1) based on the fact that an excessive measure dominated by a (measure) potential is itself a potential. For completeness we provide a new proof of this result as well.

1. Two Theorems

Let $X = (\Omega, F, F_t, \Theta_t, X_t, P^x)$ denote a right process with state space $(E, \mathcal{E})$. Here E is a U-space (see [4]) and $\mathcal{E}$ is the class of Borel sets in E. The semigroup and potential kernel of X are denoted by (P_t) and U respectively. As usual, $\Delta \notin E$ is a cemetary point to which the paths of X are banished at their lifetime $\zeta = \inf\{t: X_t = \Delta\}$. We write $\mathcal{E}^+$ for the class of positive $\mathcal{E}$-measurable functions on E.

Recall that an *excessive measure* (for (P_t)) is a σ-finite measure m on E such that $mP_t \leq m$ for all $t > 0$. One then has $mP_t \uparrow m$ as $t \downarrow 0$; see [5,(1.4)]. For example, if μ is a measure on E such that μU is σ-finite, then μU is excessive; such an excessive measure is called a *potential*. An excessive measure is *purely excessive* provided $mP_t(f) \downarrow 0$ as $t \uparrow \infty$ whenever $f \in \mathcal{E}^+$ satisfies $m(f) < \infty$. Note that any potential is purely excessive. For the representation theorem that follows, recall that an *entrance law* is a family $(\nu_t: t > 0)$ of σ-finite measures on E such that $\nu_t P_s = \nu_{t+s}$ for all s, $t > 0$.

THEOREM 1. *Let* m *be a purely excessive measure. Then there exists a unique entrance law* $(\nu_t: t>0)$ *such that* $m = \int_0^\infty \nu_t \, dt$.

Our proof of Theorem 1 relies on the following result which was proved under the present hypotheses in [5].

THEOREM 2. *Let* m *be an excessive measure and let* μU *be a potential such that* $m \leq \mu U$. *Then there exists a unique measure* ν *on* E *such that* $m = \nu U$.

REMARK. The measure ν of Theorem 2 is σ-finite: since m is σ-finite, we may choose a strictly positive $q \in E^+$ with $m(q) < \infty$. Then $Uq > 0$ and $\nu(Uq) = \nu U(q) = m(q) < \infty$.

PROOF OF THEOREM 1. It suffices to show that for each $s > 0$, there exists a unique measure ν_s on E such that $mP_s = \nu_s U$. Then for $s,t > 0$ we have

$$\nu_{t+s}U = mP_{t+s} = mP_tP_s = \nu_t UP_s = \nu_t P_s U,$$

so that $\nu_{t+s} = \nu_t P_s$ by the uniqueness of ν_{t+s}. As per the Remark above, each ν_s is automatically σ-finite. We produce ν_s by exhibiting a measure ν_s^* such that $\nu_s^* U$ is σ-finite and $mP_s \leq \nu_s^* U$; the existence and uniqueness of ν_s then follows from Theorem 2.

So fix $s > 0$ and choose $r \in]0,s[$. Then $m \geq mP_r \geq mP_s$; consequently the measure

$$\nu_s^* \equiv (s-r)^{-1}(mP_r - mP_s) \tag{3}$$

is well defined as a σ-finite measure. (Indeed, for $f \in E^+$ with $m(f) < \infty$ we take $\nu_s^*(f) = (s-r)^{-1}(mP_r(f) - mP_s(f))$. For general $f \in E^+$ there is a sequence $(f_n) \subset E^+$ with $f_n \uparrow f$ and $m(f_n) < \infty$; for such f, set $\nu_s^*(f) = \uparrow\lim_n \nu_s^*(f_n)$.) Consider now $\nu_s^* U$: pick $f \in E^+$ with $m(f) < \infty$ and for $t > 0$ compute

$\nu_s^*(\int_0^t P_u f du)$

$$= (s-r)^{-1}\int_0^t (mP_{r+u}f - mP_{s+u}f)du$$

(4)

$$= (s-r)^{-1}(\int_r^{r+t} mP_v f dv - \int_s^{s+t} mP_v f dv)$$

$$= (s-r)^{-1}(\int_r^s mP_v f dv - \int_{r+t}^{s+t} mP_v f dv).$$

But $0 \leq (s-r)^{-1}\int_{r+t}^{s+t} mP_v f dv \leq mP_{r+t}f \downarrow 0$ as $t \uparrow \infty$, since m is purely excessive. Letting $t \uparrow \infty$ in (4), we obtain

$$\nu_s^* Uf = (s-r)^{-1}\int_r^s mP_v f dv \geq mP_s f, \tag{5}$$

where the inequality follows since $s \to mP_s f$ is decreasing. We have proved (5) if $m(f) < \infty$ but (5) follows for general $f \in E^+$ by monotone convergence, since m is σ-finite. Also, by (5), we have $\nu_s^* U \leq mP_r \leq m$ so that $\nu_s^* U$ is σ-finite. Since $\nu_s^* U \geq mP_s$ as promised, we are finished.

□

PROOF OF THEOREM 2. Let m be an excessive measure and μU a potential with $m \leq \mu U$. Note that m must be purely excessive. For the moment, we shall assume that $U1 = 1$ and that $\mu(1) < \infty$. (Once the theorem is established under these auxiliary hypotheses, the general case is easily deduced.) Note that $P^{\mu}(\zeta) = \mu U1 = \mu(1) < \infty$; in particular $P^{\mu}(\zeta < \infty) = 1$. Thus X is transient and so any excessive function h admits of an approximation $Uf_n \uparrow h$ where each f_n is bounded and positive.

Arguing as in the proof of Theorem 1, if we set $\gamma_s = s^{-1}(m - mP_s)$, $s > 0$, then

(6) $$\gamma_s U = s^{-1}\int_0^s mP_v dv \uparrow m, \text{ as } s \downarrow 0.$$

Let h be an excessive function and choose (f_n) as above so that $Uf_n \uparrow h$. From (6), we see that $s \to \gamma_s Uf_n$ is decreasing; thus $s \to \gamma_s(h)$ is decreasing on $]0,+\infty[$. Similarly, if $g \le h$ are excessive functions, then $\gamma_s(g) \le \gamma_s(h)$ for all $s > 0$. Define a functional Γ by

$$\Gamma(h) = \uparrow\lim_{s\downarrow 0}\gamma_s(h), \qquad h \in H,$$

where H denotes the class of bounded excessive functions. Clearly for $g,h \in H$ and $\alpha,\beta \ge 0$,

i) $\Gamma(\alpha g+\beta h) = \alpha\Gamma(g) + \beta\Gamma(h)$;

ii) $\Gamma(g) \le \Gamma(h) \le \mu(h)$ if $g \le h$;

iii) $\Gamma(Uf) = m(f)$ if $f \in E^+$.

By ii), $\Gamma(1) \le \mu(1) < \infty$ so that $\Gamma(h) \le \|h\| \cdot \Gamma(1) < \infty$ if $h \in H$. Thus Γ extends uniquely to a bounded positive linear form on the vector lattice $D = H - H$. Let (h_n) be a sequence of positive functions from D with $h_n \downarrow 0$ pointwise. We claim that $\Gamma(h_n) \downarrow 0$ as $n \uparrow \infty$. Indeed each h_n is a difference of bounded excessive functions and so each process $(h_n(X_t): t \ge 0)$ is right continuous and left limited a.s. P^μ. Also, $h_n(X_t) = 0$ if $t \ge \zeta$ and $P^\mu(\zeta<\infty) = 1$. It now follows from the Lemme in section VII.1.2 of the second volume of [1] that

(7) $$\sup_{t\ge 0} h_n(X_t) \downarrow 0 \text{ as } n \uparrow \infty \text{ a.s. } P^\mu.$$

Fix $\varepsilon > 0$ and let $B_n = \{h_n > \varepsilon\}$. Then from (7), we deduce

(8) $$P^\mu(T(B_n)<\infty) \downarrow 0 \text{ as } n \uparrow \infty,$$

where $T(B_n) = \inf(t>0\colon X_t \varepsilon B_n)$. Let $c = \|h_1\| < \infty$. Since B_n is finely open

$$h_n(x) \leq \varepsilon + cP^x(T(B_n)<\infty), \quad \forall x \varepsilon E.$$

Thus, since $1 \varepsilon H$, $P^{\bullet}(T(B_n)<\infty) \varepsilon H$, we have

$$\begin{aligned} \Gamma(h_n) &\leq \varepsilon\Gamma(1) + c\Gamma(P^{\bullet}(T(B_n)<\infty)) \\ &\leq \varepsilon\, \mu(1) + cP^{\mu}(T(B_n)<\infty). \end{aligned} \tag{9}$$

Using (8), we let $n \uparrow \infty$ and then $\varepsilon \downarrow 0$ in (9) to obtain $\lim_n \Gamma(h_n) = 0$. Thus the positive linear form Γ satisfies the hypotheses of the Daniell extension theorem [1, III-35]: there exists a *unique* measure ν on $\sigma(\mathcal{D})$ such that $\nu(h) = \Gamma(h)$ for $h \varepsilon \mathcal{D}$. But $\sigma(\mathcal{D}) \supset E$ since X is transient, so we may (by restriction) regard ν as a measure on E. Finally, for $f \varepsilon E^+$,

$$m(f) = \Gamma(Uf) = \nu(Uf) = \nu U(f)$$

so that $m = \nu U$ as desired.

The general case can be reduced to the case already considered by employing a device of Getoor and Glover [5]. Choose $q \varepsilon E^+$ with $q > 0$ and $\mu Uq < \infty$. The absorbing set $G = \{Uq<\infty\}$ then carries each of the measures $\mu, m, \mu U$. The kernel $\tilde{U}$ defined on G by $\tilde{U}f(x) = U(qf)(x)/Uq(x)$ is the potential kernel of a right process $\tilde{X}$ on G obtained by first restricting X to G, then time changing by the CAF $t \to \int_0^t (q/Uq)(X_s)ds$, and finally h-transforming by means of $h = Uq$. Clearly $\tilde{U}1_G = 1_G$ and if we set $\tilde{m}(f) = m(qf)$, $\tilde{\mu}(f) = \mu(Uq\cdot f)$, then $\tilde{\mu}\tilde{U}(f) = \mu U(qf)$ and

$\tilde{m} \leq \tilde{\mu}\tilde{U}$. By what we have already proved, $\tilde{m} = \tilde{\nu}\tilde{U}$ for a unique measure $\tilde{\nu}$ on G. Regarding $\tilde{\nu}$ as a measure on E carried by G, we have $m = \nu U$, where $\nu(f) = \tilde{\nu}(f/Uq)$. The uniqueness of ν follows easily from that of $\tilde{\nu}$.

□

REMARKS. a) The reader familiar with [1] will note that the proof of Theorem 2 is modeled on the proof of a result to be found in VII.1 of the second volume of [1].

b) Let $L = L(\cdot,\cdot)$ denote the "energy functional" discussed in [1,XII], [8], [9] (and called the "mass" functional in [1]). For an excessive measure m and an excessive function h,

$$L(m,h) = \sup\{\mu(h): \mu U \leq m\}.$$

Using the properties of L described in [1], [8], and [9], it is easy to check that the linear form used in the proof of Theorem 2 is given by

$$\Gamma(h) = L(m,h). \tag{10}$$

In particular, a purely excessive measure m is a potential if and only if $L(m,h) = \mu(h)$, $\forall h \in \mathcal{H}$, where μ is a measure on $(E,\mathcal{E})$. In this case $m = \mu U$. A second "analytic" characterization of potentials can be obtained by scrutinizing the proof of Theorem 2. Let m be purely excessive and choose $q \in \mathcal{E}^+$ with $q > 0$ and $m(q) < \infty$. Let Γ be defined by (10) and if $f = h_1 - h_2$ where h_i is excessive with $\Gamma(h_i) < \infty$ $(i=1,2)$ then set $\Gamma(f) = \Gamma(h_1) - \Gamma(h_2)$.

Let $u = Uq$. Then m is a potential if and only if $\Gamma(h_n) \downarrow 0$ whenever (h_n) is a sequence of differences of excessive functions satisfying i) $h_n \downarrow 0$ pointwise, ii) $h_n(x) \leq c \cdot u(x)\ \forall\ x \in E$, for some constant c. In this case, Γ extends uniquely to a measure μ on $(E,\mathcal{E})$ and $m = \mu U$.

References

1. C. DELLACHERIE and P.A. MEYER. *Probabilités et Potentiel*, Vol. I, II, IV. Hermann, Paris, 1975, 1980, 1986.

2. E.B. DYNKIN. Minimal excessive measures and functions. *Trans. Am. Math. Soc. 258*, (1980), 217-244.

3. P.J. FITZSIMMONS and B. MAISONNEUVE. Excessive measures and Markov processes with random birth and death. *Probab. Th. Rel. Fields 72*, (1986), 319-336.

4. R.K. GETOOR. *Markov Processes: Ray Processes and Right Processes*. Lecture Notes in Math. *440*. Springer-Verlag, Berlin, 1975.

5. R.K. GETOOR and J. GLOVER. Markov processes with identical excessive measures. *Math. Zeit. 184*, (1983), 287-300.

6. R.K. GETOOR and J. GLOVER. Constructing Markov processes with random times of birth and death. *In this volume.*

7. R.K. GETOOR and M.J. SHARPE. Last exit times and additive functionals. *Ann. Prob. 1*, (1973), 550-569.

8. R.K. GETOOR and J. STEFFENS. The energy functional, balayage, and capacity. Manuscript (1986).

9. P.A. MEYER. Note sur l'interprétation des mesures d'équilibre. *Lecture Notes in Math. 321*, 210-216. Springer, Berlin, 1973.

P.J. FITZSIMMONS
Department of Mathematical Sciences
The University of Akron
Akron, Ohio 44325

Measures That Are Translation Invariant In One Coordinate

by

R. K. Getoor*

Let $(E,\mathcal{E})$ be a measurable space. We shall say that a measure μ on $(E,\mathcal{E})$ is Σ-finite provided μ is a countable sum of finite measures. (This notation is due to Dynkin.) Obviously any σ-finite measure is Σ-finite. It is well known that the Fubini theorem is valid for Σ-finite measures, although most text books state it only for σ-finite measures. See, for example, Theorem 7.8a in [2] for the precise statement of what we shall mean by the Fubini theorem in this note. The statement of this theorem remains valid if σ-finite is replaced by Σ-finite. It is also well known that a translation invariant Σ-finite measure μ on $\mathbf{R}$ is a multiple of Lebesgue measure ℓ; that is $\mu = c\ell$ where $0 \le c \le \infty$. For example, the proof of Theorem 60.A in [1] depends only on the Fubini theorem and so is valid for Σ-finite measures. (In [1] the much more general situation of Haar measure on a group is considered.)

The purpose of this note is to prove the following extension of the above result which seems to arise fairly often when studying stationary non-finite measures. Let $(E,\mathcal{E})$ be a measurable space and λ a Σ-finite measure on $(\mathbf{R} \times E, \mathcal{B}(\mathbf{R}) \times \mathcal{E})$. We assume that λ is *translation invariant in its first coordinate;* that is, $\lambda(F_t) = \lambda(F)$ for $t \in \mathbf{R}$ and $F \in p(\mathcal{B}(\mathbf{R}) \times \mathcal{E})$ where $F_t(s,x) = F(s + t, x)$. (If $\mathcal{F}$ is a σ-algebra $p\mathcal{F}$ denotes the positive $\mathcal{F}$ measurable functions.) For $f \in p\mathcal{E}$ it is clear that

*This research was supported in part by NSF Grant DMS 8419377.

$\varphi \to \lambda(\varphi \otimes f)$ defines a Σ-finite translation invariant measure on $\mathbf{R}$. (Here $\varphi \otimes f(s,x) = \varphi(s)f(x)$.) Consequently $\lambda(\varphi \otimes f) = \mu(f)\ell(\varphi)$ -- recall ℓ is Lebesgue measure on $\mathbf{R}$ -- and it is immediate that μ is a Σ-finite measure on E. Using this notation here is our result.

Theorem. *If* $F \in p(\mathcal{B}(\mathbf{R}) \times \mathcal{E})$, *then*

(1) $\lambda(F) = \int_{\mathbf{R}} dt \int_E F(t,x)\mu(dx) = \int_E \mu(dx) \int_{\mathbf{R}} F(t,x)dt.$

Remarks. (i) if μ is σ-finite, then (1) is just the Fubini theorem and the fact that the product measure $\ell \times \mu$ is determined by its values on functions of the form $F = \varphi \otimes f$.

(ii) If λ is not Σ-finite, then the result is false. For example, if $E = \mathbf{R}$ and μ is counting measure on E, then

$$\lambda_1(f) = \int dt \int F(t,x)\mu(dx) \quad \text{and} \quad \lambda_2(F) = \int \mu(dx) \int F(t,x)dt$$

are both translation invariant in their first coordinate but $\lambda_1 \neq \lambda_2$. Note that $\lambda_1(\varphi \otimes f) = \lambda_2(\varphi \otimes f) = \ell(\varphi)\mu(f)$. This is just the familiar counterexample to the Fubini theorem for general measures. See 7.9(b) in [2].

(iii) If λ is Σ-finite but one only assumes that $\lambda(\varphi \otimes f) = \ell(\varphi)\mu(f)$ for a Σ-finite measure μ on E (rather than the translation invariance of λ in its first coordinate), then the result is false. Let $E = \mathbf{R}$ and $\mu = \infty \cdot \ell$ and $\lambda = \infty \cdot (\ell \times \ell)$. Let $\lambda^* = \lambda + \upsilon$ where υ is one dimensional Lebesgue measure on the diagonal in $\mathbf{R}^2$. Then $\lambda \neq \lambda^*$ and one easily checks that $\lambda(\varphi \otimes f) = \ell(\varphi)\mu(f) = \lambda^*(\varphi \otimes f)$. Of course, $\mu, \lambda,$ and λ^* are Σ-finite. This example was shown to me by Klaus Janssen.

We now turn to the proof of the theorem. Since μ, ℓ, and λ are Σ-finite we may use the Fubini theorem when calculating with them. By Fubini's theorem the two integrals in (1) are equal and $\ell \times \mu$ is the measure on $\mathbf{R} \times E$ defined by their common value. Let $\lambda^* = \ell \times \mu$. The following calculation is justified by the Fubini theorem and the translation invariance of λ in its first coordinate. From now on F, G, and H denote elements of $p(\mathcal{B}(\mathbf{R}) \times \mathcal{E})$. Now for each F, G,

$$
\begin{aligned}
\lambda(F)\lambda^*(G) &= \int \mu(dx) \int dt\, \lambda(F)G(t,x) \\
&= \int \mu(dx) \int dt \left(\int F(s,y)\, \lambda(ds,dy)\right) G(t,x) \\
&= \int \mu(dx) \int dt \left(\int F(s+t,y)\, \lambda(ds,dy)\right) G(t,x) \\
&= \int \mu(dx) \int \lambda(ds,dy) \int F(s+t,y)\, G(t,x)dt \\
&= \int \mu(dx) \int \lambda(ds,dy) \int F(t,y)\, G(t-s,x)dt \\
&= \int \mu(dx) \int dt \int \lambda(ds,dy)\, F(t,y)\, G(t-s,x).
\end{aligned}
$$

But because $\lambda(\varphi \otimes f) = \ell(\varphi)\mu(f)$,

$$\int \lambda(ds,dy)F(t,y)G(t-s,x) = \int ds\, G(t-s,x) \int \mu(dy)F(t,y).$$

Therefore

$$
\begin{aligned}
\lambda(F)\lambda^*(G) &= \int \mu(dy) \int dt\, F(t,y) \int \mu(dx) \int ds\, G(t-s,x) \\
&= \lambda^*(F)\, \lambda^*(G),
\end{aligned}
$$

for all F, G. If there exists a G with $0 < \lambda^*(G) < \infty$, then $\lambda = \lambda^*$ which is the desired equality (1). If no such G exists then $\lambda^*(G) = \infty$ whenever $\lambda^*(G) > 0$. If $\lambda^* = 0$, then $\mu = 0$ and hence $\lambda = 0$. Thus we may suppose there exists a G with $\lambda^*(G) = \infty$. But then $\infty \cdot \lambda = \infty \cdot \lambda^*$. If $\lambda \neq \lambda^*$, then there exists an H with $\lambda(H) \neq \lambda^*(H)$. Since $\infty \cdot \lambda(H) = \infty \cdot \lambda^*(H)$ one must have $0 < \lambda(H) < \infty = \lambda^*(H)$. Next let $F = H/\lambda(H)$ so that $\lambda(F) = 1$, and let $g \in p\,\mathcal{B}(\mathbf{R})$ with $\ell(g) = 1$. Then calculating as before

$$1 = \ell(g)\, \lambda(F) = \int g(t) \left(\int F(s + t,x)\lambda(ds,dx)\right)dt$$
$$= \int \lambda(ds,dx) \int g(t - s)\, F(t,x)dt$$
$$= \int dt \int ds\, g(t - s) \int F(t,x)\mu(dx) = \lambda^*(F).$$

Obviously one may replace R by R^d in the above and even by a wide class of groups equipped with a left invariant Haar measure. We leave this to the interested reader.

Remark. It is easy to see that if ν and μ are non-zero Σ-finite and $\nu \times \mu$ is σ-finite, then both ν and μ are σ-finite. Thus a consequence of the theorem is that if λ is a σ-finite translation invariant in its first coordinate measure on $\mathbf{R} \times E$, then $\lambda = \ell \times \mu$ where μ is a σ-finite measure on $(E, \mathcal{E})$.

References

1. P. R. Halmos. *Measure Theory.* Van Nostrand. Toronto-New York-London. 1950.

2. W. Rudin. *Real and Complex Analysis.* McGraw-Hill, New York.

R. K. Getoor
Department of Mathematics C-012
University of California, San Diego
La Jolla, CA 92093

CONSTRUCTING MARKOV PROCESSES WITH RANDOM TIMES OF BIRTH AND DEATH

by

R. K. Getoor* and Joseph Glover**

0. Introduction

Kuznetsov [11] (see also [12]) introduced a Kolmogorov-type construction in which he constructs a stationary measure Q_m from a transition semigroup $P_t(x,dy)$ and an excessive measure m. In fact, his theorem has other interesting consequences outside of the Markovian framework, but we do not discuss these here. While Kuznetsov's proof is "elementary", it is rather involved. The purpose of this paper is to give an alternate construction of Q_m in the case of right processes. We consider both the time homogeneous and time inhomogeneous cases. Our construction does not extend to cover the other interesting cases of Kuznetsov's theorem, but our approach may yield some insight into the measures Q_m and may aid the

* Research supported by NSF Grant DMS-8419377.
** Research supported by NSF Grant DMS-8318204 and AFOSR Grant 85-0330.

reader interested in recent articles [5,10] in which the measure Q_m has played an important role. Mitro [13] has obtained a result similar to ours under duality hypotheses on the underlying processes, but her construction is quite different from ours.

We must confront squarely the complexities of the subject soon, but first we try to introduce Q_m gently to the reader by discussing the example which motivated our investigation. Let $X = (\Omega, \mathcal{F}, \mathcal{F}_t, X_t, \theta_t, P^x)$ be a right process on a Lusin state space $(E, \mathcal{E})$ with semigroup P_t and resolvent U^q. Let m be an excessive measure for X, and assume that m is in fact a measure potential. That is, $m = \mu U$ for some positive measure μ. We can "easily" construct Q_m once we introduce the measurable space on which Q_m must sit. To do this, adjoin a "birth" point a and a "death" point b to E to obtain E_b^a. Let W be the set of all maps w from $\mathbf{R}$ to E_b^a so that there is a non-empty open interval $]\alpha(w), \beta(w)[$ on which w is E-valued and right continuous, $w(t) = a$ for $t < \alpha(w)$, and $w(t) = b$ for $t > \beta(w)$. Let $Y_t(w) = w(t)$, and let $\mathcal{H}^0 = \sigma\{Y_t : t \in \mathbf{R}\}$. For each $t \in \mathbf{R}$, define a map $\rho_t : \{\zeta > 0\} \to W$ by

$$
\begin{aligned}
[\rho_t(\omega)](s) &= X_{s-t}(\omega) && \text{if } t < s \text{ and } s - t < \zeta(\omega) \\
&= a && \text{if } t > s \\
&= b && \text{if } s - t > \zeta(\omega).
\end{aligned}
$$

Let Q^t be the image of the measure P^μ under the map ρ_t: note that Q^t is a measure on $(W, \mathcal{H}^0)$. Then $Q_m = \int_{\mathbf{R}} Q^t dt$.

It is simple to check that

(0.1) $Q_m(Y_t \in dx; \alpha < t < \beta\} = m(dx)$ for every t in **R**; and

(0.2) if $t_1 < t_2 < \dots < t_n$, then

$$Q_m(\alpha < t_1, Y_{t_1} \in dx_1, \dots, Y_{t_n} \in dx_n, t_n < \beta)$$

$$= m(dx_1)P_{t_2-t_1}(x_1, dx_2)\dots P_{t_n-t_{n-1}}(x_{n-1}, dx_n).$$

The key to this construction is the fact that m is a measure potential μU. In general, excessive measures are not measure potentials, but they can be decomposed into the sum of an invariant part m^i and a potential part m^p. The potential part can be represented as an integral of an entrance law (ν_t). This proves to be enough to imitate the steps above. The representation $m^p = \int_0^\infty \nu_t dt$ is well-known, but we do not know where a direct proof of it can be found in the generality we need. In [4], Dynkin derives it as a corollary to the representation of excessive measures in terms of minimal elements. Fitzsimmons and Maisonneuve [5] have a very nice proof **using** the existence of Q_m. It is proved for finite m in [9]. In section 1, we give a direct proof of the representation of m^p (1.4). The decomposition of m is summarized in Theorem (1.10). Section 2 contains a generalization of this representation for entrance rules for a time inhomogeneous transition operator P_t^s. The main result is Theorem (2.11). Section 3 contains the construction of Q_m. In fact, we proceed more generally and construct the measure corresponding to an entrance rule and a time-inhomogeneous transition operator P_t^s.

We make the following suggestion to the reader interested only in the case of an excessive measure m and a (temporally homogeneous) right process with semigroup P_t. After reading section one, read the interpretation of the representation (1.10) given in the paragraph just below the statement of Theorem 2.33; in particular, the form (2.34) of (1.10). Then read section three with $P_t^s = P_{t-s}$ for $s < t$ and use (1.10) - that is, (2.34) - in place of (2.33) in the proof of Theorem 3.8.

We use what is essentially standard notation. Here are a few examples. Let E be a set and $\mathscr{H}$ a class of numerical functions on E. Then $b\mathscr{H}$ and $p\mathscr{H}$ denote the classes of bounded and positive functions in $\mathscr{H}$, respectively. If $(E, \mathscr{E})$ is a measurable space, then $\mathscr{E}$ is used to denote both the underlying σ-algebra and the class of all $\mathscr{E}$-measurable numerical functions on E. Thus, for example, $bp\mathscr{E} = pb\mathscr{E}$ is the class of bounded, positive, measurable functions on E. Also $\mathscr{E}^*$ denotes the σ-algebra of universally measurable sets over $(E, \mathscr{E})$. If μ is a measure on $(E, \mathscr{E})$ and $h \in p\mathscr{E}$, then $h\mu$ or $h\cdot\mu$ denotes the measure $h(x)\mu(dx)$. If $(F, \mathscr{F})$ is another measurable space and Φ is a measurable mapping from $(E, \mathscr{E})$ to $(F, \mathscr{F})$, then $\Phi(\mu)$ is the image of μ on $(F, \mathscr{F})$; that is, $\Phi(\mu)(A) = \mu[\Phi^{-1}(A)]$ for $A \in \mathscr{F}$. As usual, $\mathbf{R}$ denotes the reals and $\mathscr{B}(\mathbf{R})$ is the σ-algebra of Borel subsets of $\mathbf{R}$. Similarly, $\mathbf{R}^+$ denotes the positive (i.e. non-negative) reals and $\mathscr{B}(\mathbf{R}^+)$ the Borel σ-algebra of $\mathbf{R}^+$.

1. Excessive Measures of Right Processes.

Fix a U-space $(E, \mathscr{E})$ (i.e. E is homeomorphic to a universally measurable subset of a compact metric space), and let $X = (\Omega, \mathscr{F}, \mathscr{F}_t, X_t, \theta_t, P^x)$ be a right process on E as described in [6]. Let (P_t) and (U^q) denote the semigroup and resolvent of X, respectively.

(1.1) DEFINITION. A σ-finite measure m on $(E, \mathscr{E})$ is said to be excessive for X (or P_t or U^q) if $mP_t \leq m$ for every $t > 0$. (Here, mP_t is the measure defined by $mP_t(f) = m(P_t f)$).

It is well known that an excessive measure m also has the property that $mP_t(f)$ increases to $m(f)$ as t decreases to zero for every $f \in p\mathscr{E}$; e.g. see ([8], (1.4)).

(1.2) DEFINITION. An entrance law for X (or P_t or U^q) is a family of σ-finite measures $(\nu_t)_{t>0}$ on $(E, \mathscr{E})$ so that $\nu_t P_s = \nu_{t+s}$ for every $t > 0$ and $s > 0$.

Note that $t \to \nu_t(f)$ is $\mathscr{B}(\mathbf{R}^{++})$-measurable if $f \in p\mathscr{E}$.

The main result of this section is the theorem below connecting an excessive measure m with an entrance law ν_t. But first, we introduce the following useful convention.

(1.3) NOTATION. Let $(m_t)_{t \in \mathbf{R}}$ be a collection of σ-finite measures on $(E, \mathscr{E})$ with $m_t \geq m_{t+s}$ for every $s > 0$ and for every $t \in \mathbf{R}$. Then there exists a unique σ-finite measure π on $(E, \mathscr{E})$ so that whenever $f \in p\mathscr{E}$ with $m_s(f) < \infty$ for some

$s \in \mathbf{R}$, one has $\pi(f) = \lim_{t\to\infty} m_t(f)$. We write $\pi = \lim_{t\to\infty} m_t$. Also, if μ and ν are σ-finite measures on $(E, \mathscr{E})$ with $\mu \leq \nu$, then there exists a unique σ-finite measure λ with $\mu + \lambda = \nu$. We write $\lambda = \mu - \nu$.

(1.4) THEOREM. Let m be an excessive measure for X so that $\lim_{t\to\infty} mP_t = 0$. Then there is a unique entrance law $(\nu_t)_{t>0}$ so that $m = \int_0^\infty \nu_t dt$.

PROOF. Choose $f \in b\mathscr{E}$ with $f > 0$ and $m(f) < \infty$. Since m is excessive, $m(U^q f) < \infty$ for every $q > 0$. Let A_t be the increasing function on $]0,\infty[$ defined by setting

$$A_t = -mP_t U^q f = -e^{qt} \int_t^\infty e^{-qu} mP_u(f) du.$$

Since A_t is the product of two locally bounded absolutely continuous functions, A_t is absolutely continuous on bounded intervals. Therefore, dA_t is a positive measure on $]0,\infty[$ which is absolutely continuous with respect to Lebesgue measure. If $g \in p\mathscr{E}$, define the increasing function $A_t(g)$ on $]0,\infty[$ by setting $A_t(g) = -mP_t(g)$. Note that if $m(g) < \infty$, then

$$\int_0^\infty dA_t(g) = \lim_{r\downarrow 0} mP_r(g) - \lim_{r\to\infty} mP_r(g) = m(g).$$

In what follows, $0 < c < \infty$, and c may change from line to line. If $0 \leq g \leq cU^q f$, then $cA_t = A_t(g) + A_t(cU^q - g)$. Since both functions on the right side of this last equality are increasing, it follows that $dA_t(g) << dt$. Consequently if $g \in \mathscr{E}$ and $|g| \leq cU^q f$,

$dA_t(g)$ defines a signed measure on $]0,\infty[$ that is absolutely continuous with respect to Lebesgue measure, where $A_t(g) = -mP_t(g)$ for such g. Let $dA_t(g) = H_t(g)dt$ with g as above and with $H_t(g)$ being a finite Borel measurable version of the density. Set $E_0 = \{U^q f \geq 1\}$ and $E_n = \{(n+1)^{-1} \leq U^q f < n^{-1}\}$ for $n \geq 1$. Since $f > 0$, $E = \bigcup_{n\geq 0} E_n$. If $g \in b\mathscr{E}$, then $|g1_{E_n}| \leq cU^q f$ and so $dA_t(g1_{E_n}) = H_t(g1_{E_n})dt$. Let $(g_k) \subset p\mathscr{E}$ be a sequence increasing to $g \in b\mathscr{E}$. Then $H_t[(g - g_k)1_{E_n}] \leq H_t(g1_{E_n})$ and $\int_0^\infty H_t(g1_{E_n})dt = m(g1_{E_n}) \leq m(cU^q f) < \infty$. Hence we may apply the dominated convergence theorem to conclude

$$\int_0^\infty \lim_k H_t[(g - g_k)1_{E_n}]dt = \lim_k \int_0^\infty H_t[(g - g_k)1_{E_n}]dt$$

$$= \lim_k m[(g - g_k)1_{E_n}] = 0.$$

This shows that $H_t(g_k 1_{E_n})$ increases to $H_t(g1_{E_n})$ a.e. (dt) as k approaches infinity. By a standard result on regularizing pseudo-kernels ([7], (4.5)] or ([2], IX-11 and 13) there exists a bounded kernel $\mu_t^n(\cdot)$ from $(\mathbf{R}^+, \mathscr{B}(\mathbf{R}^+))$ to $(E, \mathscr{E})$ that is carried by E_n so that $H_t(g1_{E_n}) = \mu_t^n(g)$ a.e. (dt) for $g \in b\mathscr{E}$. Set

$$\mu_t = \Sigma_{n\geq 0}\mu_t^n.$$

Then μ_t is a kernel from $(\mathbf{R}^+, \mathscr{B}(\mathbf{R}^+))$ to $(E, \mathscr{E})$. Let $g \in pb\mathscr{E}$ with $m(g) < \infty$ and let $g_n = \Sigma_{k=0}^n g1_{E_k}$. Then

$$\int_0^\infty dA_t(g - g_n) = m(g - g_n) \to 0$$

as $n \to \infty$, and so the positive measures $dA_t(g_n)$ increase to $dA_t(g)$. Consequently

$$dA_t(g) = \mu_t(g)dt$$

for $g \in pb\mathscr{E}$ with $m(g) < \infty$ and then with one more passage to a limit for $g \in pL^1(m)$. Moreover $m = \int_0^\infty \mu_t dt$. Since on $]0,\infty[$, $\mu_{t+s}(g)dt = -d_t m(P_{t+s}g) = dA_t(P_s g) = \mu_t(P_s g)dt$ and $\mu_t(g) < \infty$ a.e. (dt) for $g \in pL^1(m)$, we have $\mu_{t+s}(g) = \mu_t(P_s g)$ a.e. (dt). But $\mathscr{E}$ is countably generated since $(E, \mathscr{E})$ is a U-space, and so for each $s > 0$

$$(1.5) \qquad \mu_{t+s} = \mu_t P_s \text{ a.e. (dt).}$$

Thus (μ_t) is a "crude" version of the desired entrance law which we shall obtain by "regularizing" (μ_t) as follows. Let L be Lebesgue measure on $\mathbf{R}^+$. Applying Fubini's theorem to (1.5), we obtain

$$\iint 1_{\{\mu_{t+s} \neq \mu_t P_s\}} ds\, dt = 0.$$

That is, there is a set $\Gamma \subset \mathbf{R}^+$ with $L(\Gamma^c) = 0$ so that for each t in Γ, there is another set $\Lambda(t) = \Lambda_t \subset R^+$ with $L(\Lambda_t^c) = 0$ and $\mu_{t+s} = \mu_t P_s$ for every s in Λ_t. Choose a sequence $(t_n) \subset \Gamma$ decreasing to zero so that $\mu_{t(n)}(U^q f) < \infty$ for each n. (This can be done since $\mu_t(U^q f) < \infty$ a.e.). For $s > 0$, define

$$(1.6) \qquad \nu^n_{t_n+s} = \mu_{t_n} P_s.$$

Note that

$$(1.7) \qquad \nu^{n}_{t_n+s} = \mu_{t_n+s} \text{ for every } s \in \Lambda(t_n),$$

and

$$(1.8) \quad \nu^{n+1}_{t_{n+1}+(t_n-t_{n+1})+s} = \mu_{t_{n+1}} P_{t_n-t_{n+1}+s}.$$

But whenever $t_n - t_{n+1} + s$ is in $\Lambda(t_{n+1})$,

$$\mu_{t_{n+1}} P_{t_n-t_{n+1}+s} = \mu_{t_n+s}.$$

Since $L(\Lambda(t_n)^c \cup \Lambda(t_{n+1})^c) = 0$, we have

$$\mu_{t_n} P_s = \mu_{t_{n+1}} P_{t_n-t_{n+1}+s} \text{ a.e. (ds)}.$$

In particular, whenever $0 \leq g \leq f$,

$$\mu_{t_n} P_s U^q g = \mu_{t_{n+1}} P_{t_n-t_{n+1}+s} U^q g \text{ a.e. (ds)}.$$

Since each side is finite and right continuous in s, they agree for all s; that is,

$$\mu_{t_n} P_s U^q = \mu_{t_{n+1}} P_{t_n-t_{n+1}+s} U^q.$$

By the uniqueness theorem for potentials ([8], (1.1)), this implies

$$\mu_{t_n} P_s = \mu_{t_{n+1}} P_{t_n-t_{n+1}+s}.$$

Thus, for every s > 0, using (1.6) and (1.8),

$$\nu^{n}_{t_n+s} = \nu^{n+1}_{t_{n+1}+(t_n-t_{n+1})+s}.$$

If $t > t_n$,

(1.9) $\nu^{n}_{t_n+(t-t_n)} = \nu^{n+1}_{t_{n+1}+(t_n-t_{n+1})+(t-t_n)} = \nu^{n+1}_{t_{n+1}+(t-t_{n+1})}$.

Define

$$\nu_t = \lim_{n\to\infty}\nu^{n}_{t_n+(t-t_n)}.$$

If $t > t_n$ and if $s > 0$, then (using (1.6)) $\nu^n_t P_s = \nu^n_{t+s}$. It follows that (ν_t) is an entrance law. By (1.7) and (1.9), $\nu_t = \mu_t$ a.e., so $m = \int_0^\infty \nu_t dt$.

To prove ν_t is unique, let γ_t be another entrance law with $m = \int\gamma_t dt$. Then

$$mP_s = \int_0^\infty \gamma_t P_s dt = \int_0^\infty \gamma_{t+s} dt = \int_0^\infty \gamma_s P_t dt = \gamma_s U.$$

Similarly, $mP_s = \nu_s U$. Since $mP_s \leq m$, and m is σ-finite, $\gamma_s = \nu_s$ by ([8], (1.1)). Q.E.D.

We can now give the representation of an excessive measure which was mentioned in the introduction.

(1.10) THEOREM. Let m be an excessive measure for X. There is a unique invariant measure m^i (i.e. $m^i P_t = m^i$ for every $t > 0$) and a unique entrance law ν_t so that

$$m = m^i + \int_0^\infty \nu_t dt.$$

This is an immediate consequence of Theorem 1.4 since it is well-known and easy to check that $m^p = \lim_{t\to\infty} mP_t$ defines an excessive measure satisfying the hypothesis of (1.4) and that $m^i = m - m^p$ defines an invariant measure. Also see the discussion following the statement of Theorem 2.33.

2. Representing an entrance rule.

Fix a U-space $(E, \mathscr{E})$. For each s and t in $\mathbf{R}$ with $s < t$ and for each x in E, let $P(s,x;\ t,dy)$ be a sub-probability measure on $(E, \mathscr{E})$. For each $s < t$, define

$$P_t^s f(x) = \int P(s,x;\ t,dy) f(y)$$

whenever $f \in b\mathscr{E}$; P_t^s is called a transition operator if (2.1), (2.2), and (2.3) are satisfied:

(2.1) $(s,t,x) \to P_t^s f(x)$ is $\mathscr{B}(\mathbf{R}) \times \mathscr{B}(\mathbf{R}) \times \mathscr{E}$ measurable for each $f \in b\mathscr{E}$.

(2.2) $P_t^s(P_u^t f) = P_u^s f$ whenever $s < t < u$ and $f \in b\mathscr{E}$.

(2.3) For each $s \in \mathbf{R}$, $P_t^s 1$ increases to 1 as t decreases to s.

We also need a type of "right" hypothesis.

(2.4) DEFINITION. A function $h_t(x)$ is called an exit rule if $h_t \in p\mathscr{E}^*$ for every t and if, for each s, $P_t^s h_t$ increases to h_s as t decreases to s.

REMARKS. (i) The argument in Dynkin ([3], Lemma (5.1)) shows that a finite exit rule must be $\mathscr{B}(\mathbf{R}) \times \mathscr{E}^*$ measurable. Standard arguments show that any entrance rule is an increasing limit of bounded entrance rules, so every entrance rule must be $\mathscr{B}(\mathbf{R}) \times \mathscr{E}^*$-measurable.

(ii) Note that h_t is an exit rule provided the function $H(t,x) = h_t(x)$ is excessive for the homogenous space-time semigroup

$$T_t((r,x);\ ds \times dy) = \varepsilon_{r+t}(ds)P(r,x;\ t + r,dy).$$

The last condition we assume for P_t^s is the following.

(2.5) For every bounded continuous function f on E and every bounded exit rule h_t, $\lim_{t\downarrow s} P_t^s(fh_t) = fh_s$.

If T_t is a right semigroup of a right process [6], then (2.5) is satisfied. In particular, if $P_t^s = P_{t-s}$ is a time-homogeneous right semigroup of a right process, then (2.5) holds.

(2.6) NOTATION. If ν is a σ-finite measure on $(E,\mathscr{E})$, then νP_t^s denotes the measure defined by $\nu P_t^s(f) = \nu(P_t^s f)$.

(2.7) DEFINITION. (i) An entrance **rule** for (P_t^s) is a family of σ-finite measures $(\nu_t)_{t\in\mathbf{R}}$ on $(E, \mathscr{E})$ so that for each t in $\mathbf{R}$, $\nu_s P_t^s$ increases to ν_t as s increases to t.

(ii) Let $-\infty < s < \infty$. An entrance **law** at s is an entrance rule (ν_t) so that $\nu_t = 0$ if $t \leq s$ and $\nu_t P_u^t = \nu_u$ whenever $s < t < u$.

(2.8) REMARK. One may apply an argument similar to that used in Dynkin ([3], Lemma (5.1)) to show that whenever (ν_t) is an entrance rule $t \to \nu_t(f)$ is $\mathcal{B}(\mathbf{R})$-measurable for $f \in p\mathcal{E}$. This result may also be obtained as a corollary of Theorem (2.33) below.

(2.9) LEMMA. _Let_ (ν_t) _be an entrance rule for_ (P_t^s). _For each_ $t \in \mathbf{R}$, _there is a function_ $f_t(x)$ _on_ E _so that_ $0 < f_t(x) \leq 1$, $\nu_t(f_t) \leq 1$ _and_ $(t,x) \to f_t(x)$ _is_ $\mathcal{B}(\mathbf{R}) \times \mathcal{E}$ _measurable_.

PROOF. For each t, ν_t is σ-finite, and we may choose an $\mathcal{E}$-measurable function k_t with $0 < k_t \leq 1$ and $\nu_t(k_t) \leq 1$. For each rational number r, choose $a_r > 0$ so that $\Sigma_{r \in \mathbf{Q}} a_r \leq 1$, and define

$$(2.10) \qquad f_t(x) = \Sigma_{r>t, r \in \mathbf{Q}} a_r P_r^t k_r(x)$$

By (2.1), $(t,x) \to f_t(x)$ is $\mathcal{B}(\mathbf{R}) \times \mathcal{E}$-measurable, and $f_t(x) \leq \Sigma_{r \in \mathbf{Q}} a_r \leq 1$. Moreover,

$$\nu_t(f_t) = \Sigma_{r>t} a_r \nu_t P_r^t k_r \leq \Sigma_{r>t} a_r \nu_r(k_r) \leq 1$$

since $\nu_r(k_r) \leq 1$. Recalling (2.3) and the fact that $k_r > 0$, we see that $P_r^t k_r(x) > 0$ for some rational $r > t$.

Therefore $f_t(x) > 0$. Q.E.D.

(2.11) THEOREM. Let (ν_t) be an entrance rule for (P_t^s) so that $\lim_{s\to-\infty}\nu_s P_t^s = 0$ for each $t \in \mathbf{R}$. Then there is a finite measure ϕ on $(\mathbf{R}, \mathscr{B}(\mathbf{R}))$ and a collection of measures $(\nu_t^s)_{s,t\in\mathbf{R}}$ so that

(2.12) for each $s \in \mathbf{R}$, $\nu^s \equiv (\nu_t^s)_{t\in\mathbf{R}}$ is an entrance law at s for (P_t^r);

(2.13) for each $f \in p\mathscr{E}$, $(s,t) \to \nu_t^s(f)$ is

$$\mathscr{B}(\mathbf{R}) \times \mathscr{B}(\mathbf{R}) \text{ - measurable;}$$

(2.14) for each $t \in \mathbf{R}$, $\nu_t = \int_{\mathbf{R}} \nu_t^s \, \phi(ds)$.

In addition, there is a strictly positive function $g_t(x)$ in $\mathscr{B}(\mathbf{R}) \times \mathscr{E}$ so that $\nu_t^s(g_t) < \infty$ for every t and s.

PROOF. Step 1: Reducing the Problem.
For $s < t$, define $Q_t^s = e^{-(t-s)}P_t^s$, and set $\mu_t = e^{-t}\nu_t$. One can easily check that Q_t^s is a transition operator and (μ_t) is an entrance rule for (Q_t^s) with $\lim_{s\to-\infty}\mu_s Q_t^s = 0$. Thus, (μ_s) and (Q_t^s) satisfy the hypotheses of the theorem and have the following extra property:

(2.15) $$\int_s^\infty Q_t^s 1 \, dt \leq \int_s^\infty e^{-(t-s)} dt = 1.$$

We now observe that it suffices to prove the theorem for (μ_s) and (Q_t^s). For suppose we can produce a family of entrance laws $(\mu^s) \equiv (\mu_t^s)$ for (Q_t^s) and a measure ϕ so that $\mu_t = \int \mu_t^s \phi(ds)$. Set $\nu_t^s = e^t \mu_t^s$. Then $\nu_t = e^t \mu_t = \int \nu_t^s \phi(ds)$. If $s < t < u$, then $\nu_t^s P_u^t = e^t \mu_t^s e^{u-t} Q_u^t = e^u \mu_u^s = \nu_u^s$. We now devote our attention to proving the theorem for (μ_s) and (Q_t^s).

Let (f_t) be the functions described in (2.9) relative to (P_t^s), and define

$$g_s = \int_s^\infty Q_t^s f_t dt. \tag{2.16}$$

Then $g_s > 0$, $(s,x) \to g_s(x)$ is $\mathscr{B}(\mathbf{R}) \times \mathscr{E}$-measurable, and, since $f_t < 1$,

$$g_s < \int_s^\infty Q_t^s 1 \; dt < 1.$$

Also, since $\nu_t(f_t) < 1$,

$$\mu_s(g_s) = \int_s^\infty \mu_s Q_t^s f_t dt < \int_s^\infty e^{-t} \nu_t(f_t) dt < e^{-s} < \infty. \tag{2.17}$$

$$\mu_t(f_t) < e^{-t}. \tag{2.18}$$

If $s < t$, then

$$Q_t^s g_t = Q_t^s \int_t^\infty Q_u^t f_u du = \int_t^\infty Q_u^s f_u du < g_s, \tag{2.19}$$

and as t decreases to s, $Q_t^s g_t$ increases to g_s. Consequently, g_t is an exit rule for (Q_t^s). If one defines $\tilde{Q}_t^s(x,dy) = g_s(x)^{-1} Q_t^s(x,dy) g_t(y)$ and $\tilde{\mu}_t = g_t \mu_t$, then $\tilde{Q}_t^s$ is a transition operator and $\tilde{\mu} = (\tilde{\mu}_t)$ is an entrance rule for $(\tilde{Q}_t^s)$ which satisfies $\tilde{\mu}_t(1) < \infty$ for all t. This additional reduction does not seem to be particularly useful in our construction, and so we shall not use it.

Step 2: Constructing ϕ.

For each t in $\mathbf{R}$ and $f \in p\mathscr{E}$, define an increasing function on $]-\infty,t[$ by setting

$$A(t,f;s) = \mu_s Q_t^s(f) \tag{2.20}$$

Note that $\lim_{s\uparrow t} A(t,f;s) = \mu_t(f)$ and $\lim_{s\to-\infty} A(t,f;s) = 0$, provided $\mu_t(f) < \infty$. If $\mu_t(f) < \infty$, then the increasing function $A(t,f;s)$ is the distribution function of the measure $\mathscr{A}(t,f)(ds) = d_s A(t,f;s)$ on $]-\infty,t[$. Note that $\mathscr{A}(t,f)(\mathbf{R}) = \mu_t(f)$. If $0 \leq f \leq g_t$, then $\mathscr{A}(t,g_t) = \mathscr{A}(t,f) + \mathscr{A}(t,g_t - f)$, so $\mathscr{A}(t,f) << \mathscr{A}(t,g_t)$. We let $\phi_t = \mathscr{A}(t,g_t)$, and we observe that ϕ_t is a finite measure on $]-\infty,t[$ since $\mu_t(g_t) \leq e^{-t}$.

Note that $\mathscr{A}(u,f_u)(\mathbf{R}) = \mu_u(f_u) \leq e^{-u}$ by (2.18) and that $u \to \mathscr{A}(u,f_u)(h)$ is $\mathscr{B}(\mathbf{R})$-measurable whenever $h \in p\mathscr{E}$. Set

$$\phi = \int_{-\infty}^{\infty} \mathscr{A}(u,f_u)du. \tag{2.21}$$

Since $\mathscr{A}(u,f_u)$ is carried by $]-\infty,u[$,

$$\phi([t,\infty[) = \int_t^\infty \mathscr{A}(u,f_u)([t,\infty[)du$$

$$\leq \int_t^\infty \mathscr{A}(u,f_u)(\mathbf{R})du \leq e^{-t}.$$

Therefore, ϕ is a Radon measure. (We shall observe at the end of the proof that ϕ can be replaced with a finite measure as promised.) From the definitions of g_t and ϕ_t, we have

$$\phi_t = (\int_t^\infty \mathscr{A}(u,f_u)du)1_{]-\infty,t[} \leq \phi.$$

Therefore, there is a function $\rho_t(s)$ so that $\phi_t(ds) = \rho_t(s)\phi(ds)$ with $0 \leq \rho_t(s) \leq 1$ and $\rho_t(s) = 0$ if $t \leq s$. Since $t \to \mu_s Q_t^s g_t$ is Borel measurable on $]s,\infty[$, ρ_t may be chosen jointly measurable in (s,t) ([2], V.T.58).

<u>Step 3: Disintegrating $\mathscr{A}(t,f)$.</u>

From Step 2 and ([2], V-T.58), we know there is a density $\alpha(t,f;s)$ which is jointly measurable in (s,t), so that

$$\mathscr{A}(t,f)(ds) = \alpha(t,f;s)\phi_t(ds)$$

whenever $0 \leq f \leq cg_t$. (Here c is any positive constant.) If $|f| \leq cg_t$, then $\mathscr{A}(t,f)(ds) = d_s\mu_s Q_t^s(f)$ defines a signed measure which is absolutely continuous with respect to ϕ_t. Hence $\mathscr{A}(t,f)(ds) = \alpha(t,f;s)\phi_t(ds)$ whenever $|f| \leq cg_t$. (The rest of this paragraph is a kernel construction analogous to the one in the first paragraph of (1.4).) Let $E_0 = \{g_t > 1\}$ and $E_n = \{(n+1)^{-1} \leq g_t < n^{-1}\}$. If $f \in p\mathscr{E}$, then

$$\mathscr{A}(t,f1_{E_n})(ds) = \alpha(t,f1_{E_n};s)\phi_t(ds)$$

since $|f| \leq \|f\|_\infty(n + 1)g_t$ on E_n. Suppose $(h_k) \subset b\mathscr{E}$ is a sequence of functions decreasing to zero. Then if $D \in \mathscr{B}(\mathbf{R})$,

$$\mathscr{A}(t,h_k 1_{E_n})(D) \leq \mathscr{A}(t,h_k 1_{E_n})(\mathbf{R}) = \mu_t(h_k 1_{E_n}),$$

and this last term goes to zero as k increases to infinity since $h_k \leq \|h_1\|_\infty$ and $\mu_t(E_n) \leq (n + 1)\mu_t(g_t) < \infty$. Consequently, if $(h_k) \subset b\mathscr{E}$ is a sequence of positive functions increasing to h, then $\alpha(t,h_k 1_{E(n)};\cdot)$ increases to $\alpha(t,h1_{E(n)};\cdot)$ a.e. (ϕ_t). By a standard result on kernels ([7],(4.5)), there exists a bounded kernel $K_t^n(s,dx)$ from $(]-\infty,t[, \mathscr{B}(]-\infty,t[)$ to $(E, \mathscr{E})$ which is carried by E_n so that for each $f \in b\mathscr{E}$, (i) $(s,t) \to K_t^n(s,f)$ is $\mathscr{B}(\mathbf{R}) \times \mathscr{B}(\mathbf{R})$-measurable, and (ii) $\alpha(t,f1_{E(n)};\cdot) = K_t^n(\cdot,f)$ a.e. (ϕ_t). Set

$$K_t(s,f) = \Sigma_{n\geq 0}K_t^n(s,f)$$

Let $f \in b\mathscr{E}$ with $\mu_t(f) < \infty$, and let

$$h_n = \Sigma_{k=0}^n 1_{E_k} f.$$

Then $\mathscr{A}(t,f - h_n)(\mathbf{R}) \leq \mu_t(f) < \infty$, and it follows that $\alpha(t,h_n;\cdot)$ increases to $\alpha(t,f;\cdot)$ a.e. (ϕ_t). Hence

$$\mathscr{A}(t,f)(ds) = K_t(s,f)\phi_t(ds)$$

whenever $\mu_t(f) < \infty$.

If $s < t < u$,

$$A(u,f;s) = \mu_s Q_u^s f = \mu_s Q_t^s Q_u^t f = A(t,Q_u^t f;s),$$

so $\mathscr{A}(u,f) = \mathscr{A}(t,Q_u^t f)$ on $]-\infty,t[$. This implies $K_u(s,f)\phi_u(ds) = K_t(s,Q_u^t f)\phi_t(ds)$ on $]-\infty,t[$ or

$$\rho_u(s)K_u(s,f)\phi(ds) = \rho_t(s)K_t(s,Q_u^t f)\phi(ds) \text{ on }]-\infty,t[.$$

Since ϕ is a Radon measure, we conclude

$$(2.22)\quad \rho_u(s)K_u(s,f) = \rho_t(s)K_t(s,Q_u^t f) \text{ a.e. } (\phi) \text{ in } s \text{ on } [-\infty,t[.$$

Set

$$(2.23)\qquad \lambda_t^s(f) = \rho_t(s)K_t(s,f)$$

Then $(s,t) \to \lambda_t^s(f)$ is $\mathscr{B}(\mathbf{R}) \times \mathscr{B}(\mathbf{R})$-measurable for every $f \in p\mathscr{E}$, and $\lambda_t^s = 0$ if $t < s$. Now (2.22) implies that for $t < u$, $\lambda_u^s(f) = \lambda_t^s(Q_u^t f)$ a.e. (ϕ) in s on $]-\infty,t[$. Since λ_u^s and λ_t^s are σ-finite measures, we have that if $t < u$,

$$(2.24)\qquad \lambda_u^s = \lambda_t^s Q_u^t \text{ a.e } (\phi) \text{ in } s \text{ on }]-\infty,t[.$$

Also,

$$\int \lambda_t^s \phi(ds) = \int_{]-\infty,t[} \rho_t(s)K_t(s,\cdot)\phi(ds)$$

$$= \int_{]-\infty,t[} K_t(s,\cdot)\phi_t(ds) = \mathscr{A}(t,\cdot)(R) = \mu_t(\cdot).$$

So (λ_t^s) is a "crude" version of the desired family (μ_t^s).

Step 4: Regularizing (λ_t^s).

By the results above and (2.17),

$$\int \lambda_t^s(g_t)\phi(ds) = \mu_t(g_t) \leq e^{-t}.$$

Thus, for each t, $\lambda_t^s(g_t) < \infty$ a.e. (ϕ). In particular, if L denotes Lebesgue measure on **R**, and if

$$\Pi = \{(r,u) : r < u, \ \lambda_u^r(g_u) < \infty, \lambda_t^r = \lambda_u^r Q_t^u$$

$$\text{a.e. L in t on } (u,\infty)\},$$

then by (2.24),

$$\int_{-\infty}^{\infty} \int_{-\infty}^{\infty} \Pi^c(r,u) 1_{\{u>r\}} \phi(dr) L(du) = 0.$$

By Fubini's theorem and the change of variables $u \to u + r$, we may rewrite this as

$$\int_{-\infty}^{\infty} \int_0^{\infty} \Pi^c(r,u + r) L(du)\phi(dr) = 0.$$

Thus there is a set $\Gamma \subset$ **R** with $\phi(\Gamma^c) = 0$ and so that if $r \in \Gamma$, then

(2.25) $\lambda_{u+r}^r(g_{u+r}) < \infty$ a.e. (L) on $\{u > 0\}$

(2.26) $\lambda_t^r = \lambda_{u+r}^r Q_t^{u+r}$ a.e. $(L \times L)$ on

$\{(t,u) : u > 0,\ t > u + r\}$.

From (2.25) and (2.26), we know there is a sequence (u_n) decreasing to zero so that

(2.27) $\lambda_t^r = \lambda_{u_n+r}^r Q_t^{u_n+r}$ a.e. (L) on $\{t : t > u_n + r\}$

(2.28) $\lambda_{u_n+r}^r (g_{u_n+r}) < \infty$.

Set $s_n = s(n) = u_n + r$, and note that s_n depends measurably on r. For each n, (2.27) implies there is a set $\Lambda_n \subset]s_n,\infty[$ of full Lebesgue measure so that

(2.29) $$\lambda_t^r = \lambda_{s_n}^r Q_t^{s_n} \text{ for all } t \in \Lambda_n.$$

Define

(2.30) $$\gamma_t^n = \lambda_{s_n}^r Q_t^{s_n} \text{ for } t > s_n.$$

Note that $\gamma_t^n = \lambda_t^r$ for every t in Λ_n. If $t > s_n > s_{n+1}$

$$\gamma_t^{n+1} = \lambda_{s_{n+1}}^r Q_t^{s_{n+1}}.$$

But if $t \in \Lambda_n \cap \Lambda_{n+1}$, then

(2.31) $$\gamma_t^n = \lambda_{s_n}^r Q_t^{s_n} = \lambda_t^r = \lambda_{s_{n+1}}^r Q_t^{s_{n+1}}.$$

Since g_t is an exit rule for Q_t^s, it follows from (2.5) that $Q_t^{s_n}(fg_t)$ and $Q_t^{s_{n+1}}(fg_t)$ are right continuous in t on $]s_n,\infty[$ whenever f is bounded and continuous. Since $L((\Lambda_n^c \cup \Lambda_{n+1}^c)\cap]s_n,\infty[) = 0$, we conclude that

$$\lambda_{s_n}^r Q_t^{s_n} = \lambda_{s_{n+1}}^r Q_t^{s_{n+1}} \text{ for every } t > s_n > s_{n+1}.$$

Thus for each $t > r \in \Gamma$, the limit

$$\mu_t^r = \lim_{n\to\infty}\lambda_{s_n}^r Q_t^{s_n} \tag{2.32}$$

exists, and for each r, (μ_t^r) is a (Q_t^s)-entrance law at r. If $r \notin \Gamma$, set $\mu_t^r = 0$ for all t. By (2.31), for every $r \in \Gamma$ and $t \in \bigcup_k \bigcap_{n>k}\Lambda_n$, $\mu_t^r = \lambda_t^r$. So $\mu_t^r = \lambda_t^r$ a.e. (L) in t on $]r,\infty[$. Let $\Phi = \{(r,t) : \mu_t^r \neq \lambda_t^r\}$. Then $\int\int \Phi(r,t)L(dt)\phi(dr) = 0$. Applying Fubini's theorem, we see there is a set $G \subset \mathbf{R}$ with $L(G^c) = 0$ so that for every t in G, $\mu_t^r = \lambda_t^r$ a.e. (ϕ). Since $\int \lambda_t^r\phi(dr) = \mu_t$, we have $\int \mu_t^r\phi(dr) = \mu_t$ for every t in G. Fix $t \in \mathbf{R}$ and choose a sequence $(t_n) \subset G$ increasing to t. Then

$$\mu_{t_n} = \int \mu_{t_n}^r \phi(dr) = \int_{]-\infty,t_n[} \mu_{t_n}^r \phi(dr),$$

so

$$\mu_{t_n} Q_t^{t_n} = \int_{]-\infty,t_n[} \mu_{t_n}^r Q_t^{t_n}\phi(dr) = \int_{]-\infty,t_n[} \mu_t^r\phi(dr).$$

Since (μ_t) is an entrance rule, $\mu_{t(n)}Q_t^{t(n)}$ increases to μ_t as t_n increases to t. The integral increases to

$$\int_{]-\infty,t[} \mu_t^r \phi(dr) = \int \mu_t^r \phi(dr),$$

and we have shown that $\mu_t = \int \mu_t^r \phi(dr)$. Now we show that $\mu_t^r(g_t) < \infty$ for every t and r. If $r \notin \Gamma$, this is clear since $\mu_t^r = 0$. Let $r \in \Gamma$ and recall that $\mu_t^r = \lambda_t^r$ a.e. (L) in t on $]r,\infty[$. By (2.25), $\mu_t^r(g_t) < \infty$ a.e. (L). In particular, there is a sequence (t(n)) decreasing to r with $\mu_{t(n)}^r(g_{t(n)}) < \infty$. But if $u > t(n)$, $\mu_u^r(g_u) = \mu_{t(n)}^r(Q_u^{t(n)} g_u) \leq \mu_{t(n)}^r(g_{t(n)}) < \infty$ since g_t is a (Q_t^s)-exit rule.

This essentially finishes the proof: we have produced the desired (μ_t^s). All that remains is to observe that we can replace ϕ with a finite measure if desired. To do this, choose a strictly positive function z on **R** so that $\phi(z) < \infty$. Set $\bar{\mu}_t^s = z(s)^{-1}\mu_t^s$ and

$$\bar{\phi}(ds) = z(s)\phi(ds) \qquad \text{Q.E.D.}$$

REMARK. Theorem (1.4) can be obtained by carefully checking through the proof of (2.11). It does not seem to be an immediate corollary of the statement of (2.11).

We can now give the representation of entrance rules.

(2.33) THEOREM. Let $\nu = (\nu_t)_{t \in \mathbf{R}}$ be an entrance rule for (P_t^s). Then for each s, $-\infty < s < \infty$, there exists an entrance law at s, $\nu^s = (\nu_t^s)$ and a finite measure ϕ on **R** so that

(i) $(s,t) \to \nu_t^s(f)$ is $\mathscr{B}(\mathbf{R}) \times \mathscr{B}(\mathbf{R})$-measurable; and

(ii) $\nu_t = \nu_t^{-\infty} + \int_{\mathbf{R}} \nu_t^s \phi(ds)$ for every $t \in \mathbf{R}$.

Before we give the proof, let us re-interpret the time homogeneous situation (1.10) in this context. If we set $\nu_t = m$ for every t, and $P_t^s = P_{t-s}$, then ν_t is an entrance rule for P_t^s. Set $\nu_t^{-\infty} = m^i$ for every t and for each $s \in \mathbf{R}$, set $\nu_t^s = \mu_{t-s}$ if $t > s$, $\nu_t^s = 0$ if $t \leq s$. In this case, we may take $\phi(dt) = dt$ to obtain

$$(2.34) \qquad m = \nu_t^{-\infty} + \int_{\mathbf{R}} \nu_t^s \phi(ds).$$

PROOF of (2.33). By definition, the measures $\nu_s P_t^s$ decrease as s decreases to $-\infty$. For each t, define $\mu_t = \lim_{s\to-\infty} \nu_s P_t^s$. Then μ_t is a σ-finite measure with $\mu_t \leq \nu_t$. Let $f \in p\mathscr{E}$ with $\nu_t(f) < \infty$; then for $s < t$,

$$\mu_s P_t^s f = \lim_{r\to-\infty} \nu_r P_s^r P_t^s f = \lim_{r\to-\infty} \nu_r P_t^r f = \mu_t(f),$$

so $\mu = (\mu_t)_{t \in \mathbf{R}}$ is an entrance law at $-\infty$. For each $t \in \mathbf{R}$, $\lambda_t = \nu_t - \mu_t$ is a σ-finite measure and $\lambda = (\lambda_t)$ is an entrance rule such that $\lim_{s\to-\infty} \lambda_s P_t^s = 0$. Apply Theorem (2.11) to λ and set $\nu^{-\infty} = \mu$ to obtain

$$\nu_t = \mu_t + \lambda_t = \nu_t^{-\infty} + \int_{\mathbf{R}} \nu_t^s \phi(ds).$$

3. Constructing the measures.

In this section, E denotes a Lusin topological space with Borel field $\mathscr{E}$ (i.e. E is homeomorphic to a Borel subset of a compact metric space). In what follows, it would suffice to assume that E is a cosouslin metrizable space, but we leave such an extension to the interested reader. Fix a transition operator (P_t^s) on $(E, \mathscr{E})$ satisfying (2.1), (2.2), (2.3) and (2.5). In order to state our last assumption on (P_t^s), we need to introduce some notation.

Let b be a point not in E, and set $E_b = E \cup \{b\}$. Topologize E_b so that E has its original topology and b is isolated in E_b. Then E_b is a Lusin topological space and the trace of its Borel field $\mathscr{E}_b$ on E is $\mathscr{E}$. We adopt the usual convention that a numerical function f on E is extended to E_b by setting $f(b) = 0$. For $-\infty < r < \infty$, let W_r denote the set of all right continuous maps from $]r,\infty[$ to E_b with b as cemetery. If $t > r$ and $w \in W_r$, let $Y_t(w) = w(t)$. Set $\mathscr{G}_r = \sigma\{Y_t : t > r\}$, and set $\beta(w) = \inf\{t : w(t) = b\}$. We now state our last assumption on (P_t^s).

(3.1) ASSUMPTION. For each $x \in E$ and $r \in \mathbf{R}$, there exists a probability $P_{x,r}$ on $(W_r, \mathscr{G}_r)$ so that if $r < t_1 < t_2 < \dots < t_n$, then

$$P_{x,r}(Y_{t_1} \in dy_1, \dots, Y_{t_n} \in dy_n, t_n < \beta)$$

$$= P_{t_1}^r(x, dy_1) P_{t_2}^{t_1}(y_1, dy_2) \dots P_{t_n}^{t_{n-1}}(y_{n-1}, dy_n)$$

(3.2) REMARKS (i) By (2.3), $P_t^s 1$ decreases to 1 as t decreases to s, so $\lim_{t \downarrow r} P'_{x,r}(t < \beta) = 1$. Thus $P_{x,r}$ is carried by $\{r < \beta\}$. It also follows from (3.1) that $x \to P_{x,r}(F)$ is $\mathscr{E}$-measurable whenever $F \in \mathscr{G}_r$.

(ii) If the space-time semigroup defined in Sec. 2 is the semigroup of a right process on $\mathbf{R} \times E$, then (3.1) holds. In particular, if $P_t^s = P_{t-s}$, where (P_t) is the semigroup of a right process on E, then (3.1) holds.

(iii) Let $r \in \mathbf{R}$, and set $W_r^+ = \{w \in W_r : \lim_{s \downarrow r} w(s)$ exists in $E_b\}$. In the usual set up for right continuous strong Markov processes, one obtains the measure $P_{x,r}$ concentrated on W_r^+. We do not need this stronger assumption here: (3.1) will suffice.

(iv) Since E_b is a Lusin space, it follows easily from IV-19 of [1] that $(W_r, \mathscr{G}_r)$ is a U-space. We need this fact below.

The usual result on constructing measures via inverse limits is stated for probabilities. Here we need a version which will work for σ-finite measures. We state it here; its proof is given at the end of this section. First, recall the definition.

(3.3) DEFINITION. Let $(F_n, \mathscr{F}_n)_{n \geq 1}$ be U-spaces and let $p_n : F_{n+1} \to F_n$ be $\mathscr{F}_{n+1}/\mathscr{F}_n$-measurable. The inverse limit $(F, \mathscr{F})$ of $(F_n, \mathscr{F}_n, p_n)$ is the subset of $\Pi_{k \geq 1} F_k$ consisting of those $x = (x_k)$ with $p_k(x_{k+1}) = x_k$ for each $k \geq 1$ and $\mathscr{F} = \sigma(q_k : k \geq 1)$, where q_k is the natural projection $q_k : (x_n) \to x_k$.

(3.4) THEOREM. Let $(F_n, \mathscr{F}_n)_{n \geq 1}$ be U-spaces and let p_n : $F_{n+1} \to F_n$ be $\mathscr{F}_{n+1}/\mathscr{F}_n$-measurable. For each n, let μ_n be a measure on $(F_n, \mathscr{F}_n)$ so that $p_n(\mu_{n+1}) = \mu_n$. Assume μ_1 is σ-finite. Then there exists a unique σ-finite measure μ on $(F, \mathscr{F})$ so that $q_n(\mu) = \mu_n$ for each $n \geq 1$.

The next result is the basic step in our construction.

(3.5) PROPOSITION. Let $\nu = (\nu_t)$ be an entrance law at r, $-\infty < r < \infty$. Then there exists a σ-finite measure Q on $(W_r, \mathscr{G}_r)$ so that $Q(\beta = r) = 0$ and if $r < t_1 < \ldots < t_n$, then

$$(3.6) \qquad Q(Y_{t_1} \in dy_1, \ldots, Y_{t_n} \in dy_n,\ t_n < \beta)$$

$$= \nu_{t_1}(dy_1) P_{t_2}^{t_1}(y_1, dy_2) \ldots P_{t_n}^{t_{n-1}}(y_{n-1}, dy_n).$$

Note that $\beta \geq r$ on W_r. The uniqueness of Q will follow from the main Theorem (3.8) of this section.

PROOF. Let (s_n) be a sequence of numbers which strictly decreases to r. For the moment, fix $k \geq 1$, and for $n \geq k$, set ${}^kW^n = W_{s(n)} \cap \{s_k < \beta\}$. Since a Borel subspace of a U-space is a U-space, it follows that ${}^kW^n$ is a U-space, and its Borel σ-algebra ${}^k\mathscr{G}^n$ is the trace of $\mathscr{G}_{s(n)}$ on ${}^kW^n$. For $n \geq k$, let $p_n : {}^kW^{n+1} \to {}^kW^n$ by restriction; that is, $p_n w(t) = w(t)$ for $t \geq s_n$. Note that the image of ${}^kW^{n+1}$, $p_n({}^kW^{n+1})$ is not all of ${}^kW^n$. In fact, it is the set $W^+_{s(n)} \cap \{s_k < \beta\}$ defined in (3.2iii). But p_n is ${}^k\mathscr{G}^{n+1}/{}^k\mathscr{G}^n$-measurable, and it is clear that $W_r \cap \{s_k < \beta\}$ may be identified with the inverse limit of

$({}^kW^n, {}^k\mathcal{G}^n, p_n)_{n>k}$. In fact, $q_n(W_r \cap \{s_k < \beta\}) = p_n({}^kW^{n+1})$, where q_n is the map from W_r to $W_{s(n)}$ defined by restriction. For each $n > k$, define ${}^kQ^n$ on $({}^kW^n, {}^k\mathcal{G}^n)$ by setting

$$(3.7) \qquad {}^kQ^n(F) = \int \nu_{s(n)}(dx)P_{x,s(n)}(F; s_k < \beta).$$

One may check that $p_n({}^kQ^{n+1}) = {}^kQ^n$ since ν is an entrance law. Let $f_k > 0$ with $\nu_{s(k)}(f_k) < \infty$. Then $f_k(Y_{s(k)}) > 0$ on ${}^kW^{k+1}$, and ${}^kQ^{k+1}(f(Y_{s(k)})) \leq \nu_{s(k)}(f_k) < \infty$. Therefore, ${}^kQ^{k+1}$ is σ-finite. Let ${}^k\mathcal{G}_r$ be the trace of $\mathcal{G}_r$ on $W_r \cap \{s_k < \beta\}$. By (3.4), there exists a σ-finite measure ${}^kQ$ on $(W_r \cap \{s_k < \beta\}, {}^k\mathcal{G}_r)$ so that $q_n({}^kQ) = {}^kQ^n$ for $n > k$. We now regard ${}^kQ$ as a measure on W_r carried by $\{s_k < \beta\}$. Set $s_0 = \infty$, and for $k \geq 1$, let ${}_kQ = 1_{\{s(k)<\beta\leq s(k-1)\}} \cdot {}^kQ$. Then set $Q = \Sigma_{k\geq 1}\, {}_kQ$. Then Q is a σ-finite measure since each term in the sum is σ-finite and they are carried by disjoint sets. If $r < t_1 < \ldots < t_n$, then

$$Q(Y_{t_1} \in dy_1, \ldots, Y_{t_n} \in dy_n, t_n < \beta)$$

$$= \Sigma_{k\geq 1}\, {}^kQ(Y_{t_1} \in dy_1, \ldots, Y_{t_n} \in dy_n, t_n < \beta \leq s_{k-1}; \beta > s_k).$$

Since the s_k decrease to r, there is an integer N so that $s_N < t_1$. Since the event ${}^kA^n = \{Y_{t_1} \in dy_1, \ldots, Y_{t_n} \in dy_n, t_n < \beta \leq s_{k-1}; \beta > s_k\} \in {}^k\mathcal{G}^N$, we may rewrite the sum as

$\Sigma_{k>1}\, {}^kQ^N({}^kA^n)$

$$= \Sigma_{k>1}\int \nu_{s(N)}(dx)P_{x,s(N)}(Y_{t_1} \in dy_1,\dots,Y_{t_n} \in dy_n,$$

$$t_n < \beta \leq s_{k-1};\ \beta > s_k)$$

$$= \int \nu_{s(N)}(dx)P_{x,s(N)}\ (Y_{t_1} \in dy_1,\dots,Y_{t_n} \in dy_n; t_n < \beta)$$

$$= \nu_{t_1}(dy_1)P_{t_2}^{t_1}(y_1,dy_2)\dots P_{t_n}^{t_{n-1}}(y_{n-1},dy_n)$$

since ν_t is an entrance law. Finally, observe that $Q(\beta = r) = 0$ because ${}_kQ$ is carried by $\{s_k < \beta\}$.

Now we come to the main result. Let a be another point not in E_b. Adjoin a as an isolated point to E_b to obtain the Lusin space E_b^a with Borel field $\mathscr{E}_b^a$. Let W be the set of all maps w from $\mathbf{R}$ to E_b^a so that there is a non-void open interval $]\alpha(w),\beta(w)[$ on which w is E-valued and right continuous, $w(t) = a$ for $t \leq \alpha(w)$ and $w(t) = b$ for $t \geq b$. (Note that for each r, $-\infty < r < \infty$, $W_r \cap \{\beta > r\}$ may be identified with $W \cap \{\alpha = r\}$). If $w \in W$, let $Y_t(w) = w(t)$, $\mathscr{H}^0 = \sigma\{Y_t : t \in \mathbf{R}\}$.

(3.8) THEOREM. _Let_ $\nu = (\nu_t)_{t\in\mathbf{R}}$ _be an entrance rule. Then there exists a unique measure_ Q_ν _on_ $(W,\mathscr{H}^0)$ _so that if_ $t_1 < \dots < t_n$,

$$(3.9)\quad Q_\nu(\alpha < t_1, Y_{t_1} \in dy_1,\dots,Y_{t_n} \in dy_n, t_n < \beta)$$

$$= \nu_{t_1}(dy_1)P_{t_2}^{t_1}(y_1,dy_2)\ \dots\ P_{t_n}^{t_{n-1}}(y_{n-1},dy_n).$$

<u>Moreover</u>, Q_ν <u>is</u> σ-<u>finite</u>.

REMARK. Note that if Q is <u>any</u> measure satisfying (3.9), then $Q = Q_\nu$ and Q is σ-finite. It is not necessary to verify that Q is σ-finite <u>a priori</u>.

PROOF. For each rational r, we may choose a decomposition of E, $(G_{rk})_{k\geq 1} \subset \mathscr{E}$ so that $\nu_r(G_{rk}) < \infty$, since ν_r is σ-finite. Let ν^s, $-\infty < s < \infty$ be the entrance laws in the representation (2.33). Recall from the statement of Theorem (2.11) and (2.9) that there is a function $h_t(x) > 0$ in $\mathscr{B}(\mathbf{R}) \times \mathscr{E}$ so that $\nu^s_t(h_t) < \infty$ for all s and t. Order the collection of sets $\{G_{rk} \cap \{\ell < h_r \leq \ell + 1\}: \ell \geq 0, k \geq 1\}$ into a sequence $(E_{rk})_{k\geq 1}$ so that $\bigcup_{k=1}^\infty E_{rk} = E$, $\nu_r(E_{rk}) < \infty$ and $\nu^s_r(E_{rk}) < \infty$ for every k and for every $s \in [-\infty,\infty[$. Let $W_{rk} = \{w \in W : Y_r(w) \in E_{rk}\}$. Since $E_{rk} \cap E_{rj} = \emptyset$ if $k \neq j$, $W_{rk} \cap W_{rj} = \emptyset$ if $k \neq j$. Because $]\alpha(w),\beta(w)[$ is non-void for each $w \in W$, one has $\bigcup_{r \in \mathbf{Q}, k\geq 1} W_{r,k} = W$.

STEP 1. <u>Uniqueness</u>.

Let Q and P be two measures on $(W,\mathscr{K}^0)$ for which (3.9) holds. Then $Q(W_{rk}) = \nu_r(E_{rk}) = P(W_{rk})$. Let Q_{rk} and P_{rk} be the restrictions of Q and P to W_{rk}. Then Q_{rk} and P_{rk} are finite measures on W_{rk}. Moreover, we have from (3.9) that $Q_{rk}(F) = P_{rk}(F)$ whenever F is of the form

(3.10) $$F = \Pi_{j=1}^n f_j \circ Y_{t_j},$$

for $t_1 \ldots < t_n$, $(f_j) \subset b\mathscr{E}$, $n > 1$. Such functions constitute a multiplication - stable vector space whose restriction to W_{rk} generates the trace of $\mathscr{H}^0$ on W_{rk}, because $\{Y_t = b\} \cap W_{rk}$ is empty if $t < r$ and $\{Y_t = a\} \cap W_{rk}$ is empty if $t > r$. Thus we have $Q_{rk} = P_{rk}$. It follows that $Q = P$.

STEP 2. Existence.

Let Q^s be the σ-finite measure on $(W_s, \mathscr{G}_s)$ carried by $\{s < \beta\}$ constructed from ν^s in Proposition (3.5). For each s, $-\infty < s < \infty$, define maps q_s: $W_s \cap \{s < \beta\} \rightarrow W$ by setting

$$q_s w(t) = w(t) \text{ if } t > s$$

$$= a \quad \text{ if } t \leq s.$$

Note that $\alpha \circ q_s = s$ and $\beta \circ q_s = \beta$. Let ${}^sQ = q_s(Q^s)$. Then ${}^sQ$ is a measure on $(W, \mathscr{H}^0)$ carried by $\{\alpha = s\}$. If $t_1 < \ldots < t_n$, then from (3.6) we have

$$(3.11) \quad {}^sQ(Y_{t_1} \in dy_1, \ldots, Y_{t_n} \in dy_n, t_n < \beta)$$

$$= \nu^s_{t_1}(dy_1) P^{t_1}_{t_2}(y_1, dy_2) \ldots P^{t_{n-1}}_{t_n}(y_{n-1}, dy_n),$$

and note that this is zero if $t_1 \leq s$. In particular, ${}^sQ(W_{rk}) = \nu^s_r(E_{rk}) < \infty$. Thus each ${}^sQ$ is σ-finite.

Next, we claim that $s \rightarrow {}^sQ(F)$ is Borel measurable for each $F \in \mathscr{H}^0$. In view of the above and by "disjointing" the W_{rk}, it suffices to prove this for each $s \rightarrow {}^sQ(F \cap W_{rk})$

with F of the form (3.10). But for such F,

$$(3.12)\quad {}^{s}Q(F) = \int f_1(x_1)P_{t_2}^{t_1} f_2 P_{t_3}^{t_2} f_3 \ldots P_{t_n}^{t_{n-1}} f_n(x_1)\nu_{t_1}^{s}(dx_1).$$

Note that this expression is zero if $s > t_1$ as it should be since $\alpha = s$ almost surely ${}^{s}Q$, and that it is Borel measurable in s. Consequently, so is $s \to {}^{s}Q(F)$ for all $F \in \mathscr{H}^0$. Finally, define Q_ν on $\mathscr{H}^0$ by setting

$$(3.13)\qquad Q_\nu(F) = {}^{-\infty}Q(F) + \int_{\mathbf{R}} {}^{s}Q(F)\phi(ds)$$

for $F \in \mathscr{H}^0$. Using (3.12) and (2.33), it is immediate that (3.9) holds for Q_ν. Moreover, $Q_\nu(W_{rk}) = \nu_r(E_{rk}) < \infty$, so Q_ν is σ-finite. This establishes the existence of Q_ν. Q.E.D.

PROOF OF THEOREM (3.4).

Let h_1 be finite and strictly positive on F_1 with $\mu_1(h_1) = 1$. Define inductively $h_{n+1} = h_n \circ p_n$ for $n > 1$. Then $h_n \in \mathscr{F}_n$, $h_n > 0$ on F_n, and $\mu_n(h_n) = \mu_1(h_1) = 1$ for each n. Let $\nu_n = h_n\mu_n$. Then ν_n is a probability on $(F_n, \mathscr{F}_n)$, and one easily checks that $p_n(\nu_{n+1}) = \nu_n$. Hence by (III-53, [1]), there exists a unique probability ν on $(F, \mathscr{F})$ with $\nu_n = q_n(\nu)$. Define h on F by $h(x) = h_n(x_n) = h \circ q_n(x)$. (Here, $x = (x_k)$). Since $h_{n+1}(x_{n+1}) = h_n \circ p_n(x_{n+1}) = h_n(x_n)$, h is well-defined, $h \in \mathscr{F}$, and $h > 0$. Let $\mu = h^{-1}\nu$. Then μ is σ-finite, and one readily checks that $q_n(\mu) = \mu_n$. Finally, the uniqueness of μ follows from the uniqueness of $\nu = h\mu$. Q.E.D.

(3.14) REMARK. Let us restrict ourselves to the temporally homogeneous case. The assumption that E is a Lusin (or at least cosouslin) metrizable space is critical for our construction since this is what allows us to use (3.4) in the proof of (3.5). However, the assumption that P_t maps $b\underline{\underline{E}}$ into $b\underline{\underline{E}}$ is not needed; it suffices that P_t maps $b\underline{\underline{E}}$ into $b\underline{\underline{E}}^*$. Thus one has the following statement: let (P_t) be a right (not necessarily Borel) semigroup on a Lusin topological space $(E, \underline{\underline{E}})$. If $v = (v_t)$ is an entrance rule for P_t, then there exists a unique measure Q_v on $(W, \underline{\underline{H}}^\circ)$ satisfying (3.9) with $P_t^s = P_{t-s}$ for s smaller than t. In this situation, the function $h_t(x)$ and sets E_{rk} in the proof of (3.8) are only $\underline{\underline{E}}^*$ measurable. But this causes no difficulty.

References

1. C. Dellacherie and P. A. Meyer. Probabilities and Potential. North-Holland. Amsterdam-New York-Oxford. 1978.

2. C. Dellacherie et P. A. Meyer. Probabilités et Potentiel. Chap. V-VIII et Chap. IX-XI. Hermann. Paris. 1980 and 1983.

3. E. B. Dynkin. Regular Markov processes. Russian Math. Surveys. 28 (1973) 33-64. Reprinted in London Math. Soc. Lecture Note Series 54. Cambridge Univ. Press. 1982.

4. E. B. Dynkin. Minimal excessive measures and functions. Trans. Amer. Math. Soc. 258 (1980) 217-244.

5. P. J. Fitzsimmons and B. Maisonneuve. Excessive measures and Markov processes with random birth and death. To appear in Z. Wahrscheinlichkeitstheorie verw. Geb.

6. R. K. Getoor. Markov Processes: Ray Processes and Right Processes. Lecture Notes in Math. 440. Springer. Berlin-Heidelberg-New York. 1975.

7. R. K. Getoor. On the construction of kernels. Sém. de Prob. IX. Lecture Notes in Math. 465, 443-463. Springer. Berlin-Heidelberg-New York. 1975.

8. R. K. Getoor and J. Glover. Markov processes with identical excessive measures. Math. Zeit. 184 (1983) 287-300.

9. R. K. Getoor and M. J. Sharpe. Last exit times and additive functionals. Ann. Prob. 1 (1973) 550-569.

10. R. K. Getoor and J. Steffens. Capacity theory without duality. Submitted to Z. Wahrscheinlichkeitstheorie verw. Geb.

11. S. E. Kuznetsov. Construction of Markov processes with random times of birth and death. Theory Prob. and Appl. 18 (1974) 571-575.

12. S. E. Kuznetsov. Nonhomogeneous Markov processes. Journal of Soviet Math. 25 (1984) 1380-1498.

13. J. B. Mitro. Dual Markov processes: construction of a useful auxiliary process. Z. Wahrscheinlichkeitstheorie verw. Geb. 47 (1979) 139-156.

14. M. J. Sharpe. General Theory of Markov Processes. Forthcoming book.

R. K. Getoor
Department of Mathematics, C-012
University of California
La Jolla, CA 92093

J. Glover
Department of Mathematics
University of Florida
Gainesville, FL 32611

Branching Brownian Motion and the Dirichlet Problem of a Nonlinear Equation

*Pei Hsu**

§1. Introduction

We consider a simple case of Markov branching processes. Suppose we are given the following data:

(i) A probability vector $F = \{p_2, p_3, \ldots\}, p_i \geq 0$ and $\sum_{i=2}^{\infty} p_i = 1$.

(ii) A nonnegative measurable function b on R^d.

Then a (b, F)-branching Brownian motion on R^d can be described as follows. At a point $x \in R^d$, start an ordinary Brownian motion B. Choose a random time T obeying the law

$$P[T > t | B] = e(t) \stackrel{\text{def}}{=} e^{-\int_0^t b(B_s)ds} \tag{1.1}$$

and an integral random variable M obeying the law

$$P[M = n | T, B] = p_n. \tag{1.2}$$

At time T, the Brownian particle splits into M independent particles and these particles start their own lives according to the law we have just described. The stochastic process (stochastic shower) $X = \{X_t; t \geq 0\}$ thus obtained has the strong Markov property interpreted in the obvious way. Note that X_t now stands for a finite or infinite particles moving randomly in R^d and we write

* *Research supported in part by the grant NSF-MCS-82-01599.*

$X_t = \left\{X_t^{(1)}, X_t^{(2)}, \ldots, X_t^{(m(t))}\right\}$, where $m(t)$ is the number of particles at time t. Given a function f on R^d and a finite set $S \in R^d$, the symbol $f^*(S)$ stands for the product of the values of f on S. Thus if $m(t)$ is finite,

$$f^*(X_t) = \prod_{i=1}^{m(t)} f\left(X_t^{(i)}\right)$$

Now suppose $||f||_\infty \leq 1$, we consider the expression

$$u(t, x) = E^x\left[f^*(X_t); m(t) < \infty\right]. \tag{1.3}$$

It can be shown easily by the Markov property that function $u(t, x)$ is the solution of the nonlinear parabolic equation:

$$\frac{\partial u}{\partial t} = \frac{1}{2}\Delta u + b[F(u) - u], \qquad u(0, \cdot) = f \tag{1.4}$$

where

$$F(u) = \sum_{i=2}^{\infty} p_n u^n.$$

Thus it is natural to use the (b, F)-branching Brownian motion to discuss the corresponding Dirichlet problem:

$$\begin{cases} \frac{1}{2}\Delta u + b[F(u) - u] = 0, & \text{on } \Omega \\ u = f, & \text{on } \partial\Omega \end{cases} \tag{1.5}$$

where Ω is a bounded domain in R^d. To explore this connection is the main purpose of the present note. We will denote the boundary value problem (1.5) by $D(\Omega, F, b; f)$.

The existence of solution depends on the magnitude of the boundary function. Probabilistically it depends on the speed the Brownian particles accumulate on the boundary. Our discussion centers on the problem of validity of the expression

$$u_f(x) = E_x[f^*(X_{\tau_\Omega}); N < \infty] \tag{1.6}$$

as a solution to the problem (1.5). Let us explain the notation used in (1.6). A particle of the branching Brownian motion will almost surely hit the boundary.

We imagine that each particle is stopped at the first time it hits the boundary. Thus eventually either the process of branching inside Ω ceases at a finite time or this process will go on forever. In the former case there are only finitely many particles ending up on the boundary, whereas in the latter case the number of points on the boundary goes to infinity with time. For a fixed time, let N_t be the number of particles which have already reached the boundary before time t and let $N = \lim_{t\to\infty} N_t$. The symbol X_{τ_Ω} denotes the set of positions of the particles which eventually reach the boundary. Thus X_{τ_Ω} is finite set on $\{N < \infty\}$ and (1.6) has a meaning (See [2] for an extensive discussion of Markov branching processes).

As observed in [4], for $\|f\|_\infty \leq 1$, the function u defined in (1.6) is always a solution of $D(\Omega, F, b; f)$. We will show that u_f may represent a solution to the boundary value problem even when $\|f\|_\infty > 1$. How large boundary functions can be allowed depends on the domain Ω and the branching rate b. The smaller the domain in area and the smaller the function b (the slower the branching speed), the larger the boundary function can be allowed (Theorem 3.2). The existence depends essentially on the convergence of the expression (1.6). Let $q_n(x) = P_x[N = n]$. Then the problem can be solved for large boundary functions if the probabilities $q_n(x)$ decreases to zero at least as fast as a geometric progression with a small ratio. In the reverse direction, we show that for any domain, the problem cannot be solved if the boundary function is too large. This requires to show that $q_n(x)$ no faster than a geometric progression.

In §4, we deal with the case $\|f\|_\infty \leq 1$. In this case the uniqueness problem can be completely settled. [5] contains a discussion of this case for constant branching rate b. Our argument based on the martingale theory is more probabilistic.

Now a few words about basic assumptions in this note. We always assume Ω is a domain with finite area. To simplify the discussion we assume that the radius of convergence of F is infinite. We assume there exist two constants c_1

and c_2 such that $0 < c_1 \leq b \leq c_2$. Without assuming any smoothness on the data, by a solution of the problem $D(\Omega, F, b; f)$ we mean a function u which is continuous on $\bar{\Omega}$ and satisfies

$$u = G_\Omega[bF(u); b] + H_\Omega(f; b) \tag{1.7}$$

where

$$G_\Omega[v; b] = -\left[\frac{\Delta}{2} - b\right]^{-1} v = E_\cdot\left[\int_0^{\tau_\Omega} e(s)v(B_s)ds\right]$$

and

$$H_\Omega(f; b) = E_\cdot\left[e(\tau_\Omega)f(B_{\tau_\Omega})\right].$$

A solution in this sense is classical if the data are sufficiently smooth; see the discussion in [5]. To simplify notation we often write $\tau = \tau_\Omega$.

As in [5], the method used here can be applied to Markov branching diffusion of more general type and to the case where F may depend on the space variables and may take both signs. See also [3] for discussions of related problems from a different point of view.

§2. Basic Representation Theorem

Under our assumption on F, we have $F'(1) < \infty$. Therefore $m(t) < \infty$ a.s. for any finite t. Furthermore, for any bounded function f, there exists an $\epsilon > 0$ such that $f^*(X_t)$ is integrable for $0 \leq t \leq \epsilon$. Hence $u(t, x)$ in (1.3) is well defined as least for small time. These facts can be proved by using formula (7) of [1], p.106.

Let u_f be defined as in (1.6). For a nonnegative f, function u_f is always defined but may be infinite.

Proposition 2.1. *Suppose that f is bounded or nonnegative. Then u_f satisfies (1.7).*

Proof. This is a simple application of the strong Markov property. Let $\tau = \tau_\Omega$ as before. We always use B to denote the base Brownian motion of X. Recall that T is the first splitting time. Using the Markov property at $\tau \wedge T$, we have by (1.2)

$$\begin{aligned} u_f(x) =& E_x\left[E^*_{X_{\tau\wedge T}}\left[f^*(X_\tau); N < \infty\right]\right] \\ =& E_x\left[f(B_\tau); \tau < T\right] + E_x\left[E^*_{X_T}\left[f^*(X_\tau); N < \infty\right]; \tau \geq T\right] \\ =& E_x\left[f(B_\tau); \tau < T\right] + E_x\left[F(u(B_T)); \tau \geq T\right]. \end{aligned}$$

(1.7) follows from this identity and (1.1).

Before proving the next proposition, we need a lemma.

Lemma 2.2. *Let* $u(t,x) = E_x\left[u^*(B_{\tau_\Omega \wedge t})\right]$. *If* u *is bounded and satisfies (1.7), then* $u(t,x) = u(x)$.

Proof. Note that *a priori* we do not know the random variable $u^*(X_{\tau\wedge t})$ is integrable. But from the remark at the beginning of this section we know it is so for small t. Thus by the semigroup property, it is sufficient to establish the result for small t. Split the integral $E_x\left[u^*(X_{\tau\wedge t})\right]$ into three pieces: $[\tau \leq T]$, $[t \leq T < \tau]$ and $[T < \tau \wedge t]$. Using (1.1) and (1.2) and the Markov property,

$$\begin{aligned} u(t,x) =& E_x\left[u(B_{\tau\wedge t})e(\tau)\right] + E_x\left[u(B_t)(e(t) - e(\tau)); t < \tau\right] \\ &+ E_x\left[F(u(t-T, B_T); T < \tau \wedge t\right] \\ =& E^x\left[u(B_{\tau\wedge t})e(\tau \wedge t)\right] \\ &+ E_x\left[\int_0^{\tau\wedge t} F(u(t-s, B_s))e(s)b(B_s)ds\right]. \end{aligned} \tag{2.1}$$

On the other hand, from (1.7) we have

$$u(x) = E_x\left[u(B_{\tau\wedge t})e(\tau \wedge t)\right] + E_x\left[\int_0^{\tau\wedge t} F(u(B_s)e(s)b(B_s)ds\right]. \tag{2.2}$$

Subtracting (2.2) from (2.1) and using $|F(u) - F(v)| \leq K|u - v|$, we obtain

$$\epsilon(t,x) \leq K \int_0^t E_x\left[\epsilon(t-s, B_s); s < \tau\right] ds$$

with $\epsilon(t,x) = |u(t,x) - u(x)|$. Integrating over Ω and using Gromwall's inequality, we see $\epsilon(t,x) \equiv 0$. The lemma is proved.

Theorem 2.3. *Suppose u is a solution of $D(\Omega, F, b; f)$. Then*

$$\left\{M_t \stackrel{\text{def}}{=} u^*(B_{\tau_\Omega \wedge t}); t \geq 0\right\}$$

is a P_x-martingale for any $x \in \Omega$.

Proof. By Proposition 2.1 and Lemma 2.2, the random variable M_t is integrable. Now for any $s \leq t$,

$$E_x\left[M_t \middle| \mathcal{F}_{\tau\wedge s}\right] = E^*_{X_{\tau\wedge s}}\left[M_{\tau\wedge(t-s)}\right] = u^*(X_{\tau\wedge s}) = M_s.$$

Therefore, $M = \{M_t; t \geq 0\}$ is a $\mathcal{F}_{\tau\wedge t}$-martingale.

Proposition 2.4. *(Minimality of the probabilistic solution) Let $f \geq 0$ and let u be a solution of $D(\Omega, F, b; f)$. Let u_f be the probabilistic solution (1.6). Then $0 \leq u_f \leq u$.*

Proof. Follows easily from the preceding proposition. We have

$$\begin{aligned} u(x) =& E_x\left[u^*(X_{\tau\wedge t})\right] \geq \lim_{t\to\infty} E_x\left[u^*(X_{\tau\wedge t}); N < \infty\right] \\ \geq& E_x\left[\lim_{t\to\infty} u^*(X_{\tau\wedge t}); N < \infty\right] \geq E_x\left[u^*(X_\tau); N < \infty\right] \\ =& u_f(x). \end{aligned}$$

§3. Existence of Solutions

In view of Proposition 2.1, we look for conditions on the boundary function under which the expression (1.6) is meaningful. We use a very simple-minded estimate:

$$|u_f(x)| \leq \sum_{n=1}^{\infty} \|f\|_\infty^n q_n(x). \tag{3.1}$$

Here $q_n(x) = P_x[N = n]$. Thus the boundedness of u_f depends on the decreasing rate of the probabilities q_n. Letting $f \equiv \alpha$ in (1.7) and comparing coefficients for powers of α, we see that q_n satisfies the following recursion formula:

$$q_n = G_\Omega[bH_n(q_1, \cdots, q_{n-1}); b] \tag{3.2}$$

where H_n's are determined by

$$F\left(\sum_{n=1}^{\infty} a_n \xi^n\right) = \sum_{n=2}^{\infty} H_n(a_1, \cdots, a_{n-1})\xi^n,$$

and

$$q_1 = E_\cdot\left[e(\tau_\Omega)\right].$$

We say that F terminates at m_0 if $p_n = 0$ for $n > m_0$. We need the following simple lemma, whose proof we omit.

Lemma 3.1. *(a) Assume that F terminates. Let sequence $A_n, n \geq 1$ be defined by $A_1 = 1$ and $A_n = H_n(A_1, \ldots, A_{n-1})$. Then the power series $\sum_{n=1}^{\infty} A_n \xi^n$ has a positive radius of convergence $r \geq 3 - 2\sqrt{2}$. (b) Let I_n be defined by*

$$p_{n_0}\left(\sum_{n=1}^{\infty} a_n \xi^n\right) = \sum_{n=2}^{\infty} I_n(a_1, \cdots, a_{n-1})\xi^n.$$

Let $c > 0$. Let sequence $B_n, n \geq 1$ and let defined by $B_1 = c$ and $B_n = I_n(B_1, \ldots, B_{n-1})$. Then the power series $\sum_{n=1}^{\infty} B_n \xi^n$ has a finite radius of convergence.

The following result gives a lower and an upper bound for the probabilities $q_n(x)$.

Proposition 3.2. *(a) Assume that F terminates at m_0. There exist a K_1 and a positive γ independent of F such that for all $n \geq 1$*

$$\|q_n\|_\infty \leq K_1 \left(\gamma \|b\|_\infty |\Omega|^{2/d}\right)^{\frac{n-1}{m_0-1}}. \tag{3.3}$$

(b) Assume Ω is bounded and smooth. There exist positive constants K_2 and β such that for all $n \geq 1$

$$q_n(x) \geq K_2 \beta^{n-1} h(x). \tag{3.4}$$

Here

$$h = G_\Omega[b; b] = 1 - E_\cdot\left[e(\tau_\Omega)\right].$$

Proof. (a) Upper bound. Define A_n as in the preceding lemma. We prove by induction that

(3.5) $$\|q_n\|_\infty \le A_n \|h\|_\infty^{\frac{n-1}{m_0-1}}.$$

The inequality holds obviously for $n = 1$. To go from $n-1$ to n, we have by (3.2) and the hypothesis that F terminates at m_0.

$$\begin{aligned} u_n =& G_\Omega[H_n(q_1, \ldots, q_{n-1})b; b] \\ \le& \|h\|_\infty^{\frac{n-m_0}{m_0-1}} H_n(A_1, \ldots, A_{n-1}) G_\Omega[b; b] \\ \le& A_n \|h\|_\infty^{\frac{n-1}{m_0-1}}. \end{aligned}$$

This proves (3.5). Now by the preceding lemma there are positive constants K_1 and γ_1 such that $A_n \le K_1\gamma_1^n$. On the other hand we have for some constant universal constant γ_2

$$\|h\|_\infty \le E_x\left[\int_0^{\tau_\Omega} e(s)b(B_s)ds\right] \le \|b\|_\infty E_x[\tau_\Omega] \le \gamma_2\|b\|_\infty|\Omega|^{2/d}.$$

(3.3) follows with $\gamma = \gamma_1\gamma_2$.

(b) Lower bound. Take $c = \min_{x\in\bar\Omega} E_x[e(\tau_\Omega)] > 0$ and define B_n as in the preceding lemma. We prove by induction that

(3.6) $$q_n(x) \ge B_n \beta_1^{n-1} h(x)$$

where

$$\beta_1 = \min_{x\in\bar\Omega} \frac{G_\Omega[h^{n_0}b; b](x)}{h(x)}.$$

$\beta_1 > 0$ because by the smoothness of the domain, both $G_\Omega[h^{n_0}b; b]$ and h vanish on the boundary exactly to the first order. Now (3.6) holds for $n = 1$ by the definition of c, since $q_1(x) = E_x[e(\tau_\Omega)]$. For the induction step, we have

$$\begin{aligned} q_n \ge& G_\Omega[H_n(q_1, \ldots, q_{n-1})b; b] \\ \ge& \beta_1^{n-2} I_n(B_1, \ldots, B_{n-1}) G_\Omega[h^{n_0}b; b] \\ \ge& B_n\beta_1^{n-1}h \end{aligned}$$

Now by part (b) of the preceding lemma, there exist K_2 and β_2 such that $B_n \ge K_2\beta_2^n$. (3.4) follows with $\beta = \beta_1\beta_2$.

The following results follow immediately from the lower and upper bounds and Proposition 2.1 and 2.4.

Theorem 3.3. *(i) Assume that F terminates at m_0. There exists a constant γ independent of F such that the problem $D(\Omega, F, b; f)$ has a solution if*

$$\|f\|_\infty < \left(\gamma\|b\|_\infty|\Omega|^{2/d}\right)^{-\frac{1}{m_0-1}}.$$

(ii) Assume Ω is bounded and smooth. There exists a constant $\beta = \beta(\Omega, F, b)$ such that the problem has a solution if $\|f\|_\infty < \beta$ and has no positive bounded solution if $f > \beta$.

§4. The case $\|f\|_\infty \leq 1$

In this section a solution means a solution with $\|u\|_\infty \leq 1$. Define

$$\lambda = \lambda_1(b, \Omega) \stackrel{\text{def}}{=} \inf_{\substack{v \\ v|\partial\Omega=0}} \frac{\int_\Omega |\nabla v|^2}{\int_\Omega bv^2}. \tag{4.1}$$

The infimum is attained by a positive continuous function ϕ vanishing on the boundary in the sense that for any continuous v vanishing on the boundary

$$\int_\Omega bv^2 \geq \frac{\lambda}{2}\int_\Omega bvG_\Omega(bv) \tag{4.2}$$

and

$$\phi = \frac{\lambda}{2}G_\Omega(b\phi). \tag{4.3}$$

Here $G_\Omega = (-\Delta/2)^{-1}$ is the Green operator of Ω with Dirichlet boundary condition. Let

$$\alpha = F'(1) - 1 - \frac{1}{2}\lambda_1(b, \Omega).$$

The following results can be established. Let

$$A(t, E) = \text{ the number of particles in } E \subset \Omega \text{ at time } t.$$

We have

$$E_x\left[A(t, E)\right] = e^{\beta t}\psi(x)\int_E \psi + o(e^{\beta t}) \tag{4.4}$$

where β is the first eigenvalue of $\Delta/2 + [F'(1) - 1]b$:

$$\beta = - \inf_{v|_{\partial\Omega}=0} \frac{\frac{1}{2}\int_\Omega |\nabla v|^2 - [F'(1) - 1]\int_\Omega bv^2}{\int_\Omega v^2} \tag{4.5}$$

and ψ is its normalized eigenfunction. Furthermore, let

$$\mu(E) = \frac{\int_E \psi}{\int_\Omega \psi}.$$

Then there is a sequence $t_n \to \infty$ such that for any $x \in \Omega$

$$P_x \left[\frac{A(t_n, E)}{A(t_n, \Omega)} \to \mu(E) \middle| N = \infty \right] = 1. \tag{4.6}$$

Both (4.4) and (4.6) can be proved by eigen-expansion, starting from (1.4) (cf.[5]). We also notice that α and β always take the same sign. This is an immediate consequence of the variational characterizations (4.1) and (4.5), the definition of α and the lower bound of b.

Theorem 4.1. *Suppose u is a solution of $D(\Omega, F, b; f)$, $\|u\|_\infty \leq 1$ but $u \not\equiv 1$. Then it is the unique solution with such property.*

Proof. By the assumption, there must be a set E of positive measure on which $|u| \leq \epsilon < 1$. From Theorem 2.3,

$$u(x) = E_x [u(B_{\tau\wedge t}); N < \infty] + E_x [u(X_{\tau\wedge t}); N = \infty]. \tag{4.7}$$

If $P_x [N = \infty] = 0$, the second term on the right side is zero. Otherwise using (4.6), we obtain

$$\left| E_x [u(B_{\tau\wedge t_n}); N = \infty] \right| \leq E_x \left[\epsilon^{A(t_n, E)}; N = \infty \right] \to 0.$$

The uniqueness follows then by taking limit in (4.7).

It follows from the theorem that the boundary value problem has exactly one solution if $\|f\|_\infty \leq 1$ but $f \not\equiv 1$ and at most two solutions if $f \equiv 1$. The case $f \equiv 1$ is critical because 1 is the root of $F(u) - u = 0$. The case $b =$ const. of the following result was discussed in [5].

Theorem 4.2. *If $\alpha \leq 0$ then $u \equiv 1$ is the unique solution of $D(\Omega, F, b; 1)$, and $N < \infty$ a.s. If $\alpha > 0$ then the extinction probability*

$$u_1(x) = P_x[N < \infty]$$

is the only other solution of the problem and $0 < u_1 < 1$.

Proof. (a) $\alpha \leq 0$. Assume u is a solution. Let $v = 1 - u$. If $v \not\equiv 0$, we have by (1.7) and $F(1-v) - 1 > vF'(1)$,

$$v = G_\Omega[b(1-v) - bF(1-v)] < -F'(1)G_\Omega(bv).$$

Multiply both sides by v and integrate. Using (4.2), we have

$$\alpha \int_\Omega bv^2 > 0,$$

a contradiction. Therefore $v \equiv 0$.

Here is an alternative proof. By (4.7), it is enough to show $P_x[N < \infty] = 1$ for all $x \in \Omega$. As mentioned above, $\alpha \leq 0$ is equivalent to $\beta \leq 0$. Thus by (4.4) we see that $E_x[A(t, \Omega)] \leq M$ for a constant M. On the other hand, since b is bounded from above, we can show that with probability one either $A(t, \Omega) \to 0$ or ∞ as $t \to \infty$ (cf.[4]). Consequently we must have $A(t, \Omega) \to 0$, a.s., which is equivalent to $N < \infty$, a.s.

(b) $\alpha > 0$. By Theorem 4.1, we need only prove the assertion $0 < u_1 < 1$. Let $v = 1 - \epsilon\phi$, where ϕ is the first eigenfunction; see (4.3). $u_1 < 1$ is implied by the assertion that

$$\left\{Q_t \stackrel{\text{def}}{=} v^*(X_{\tau_\Omega \wedge t}); t \geq 0\right\}$$

is a P_x-supermartingale for small ϵ. For then we have

$$\begin{aligned} 1 > & v(x) \geq E_x[v(X_{\tau \wedge t})] \geq E_x\left[\lim_{t \to \infty} v(X_{\tau \wedge t}); N < \infty\right] \\ = & P_x[N < \infty] = u_1(x). \end{aligned}$$

To show that Q is a supermartingale, all we need is

$$v(t, x) \stackrel{\text{def}}{=} E_x[v(X_{\tau \wedge t})] \leq v(x). \tag{4.8}$$

We have as in (2.1)

$$v(t,x) = E_x\left[v(B_{\tau\wedge t})e(\tau\wedge t)\right] + E_x\left[\int_0^{\tau\wedge t} F(v(t-s,B_s))b(B_s)e(s)ds\right]. \tag{4.9}$$

But instead of (2.2), we have by (4.3)

$$v(x) = E_x\left[v(B_{\tau\wedge t})e(\tau\wedge t)\right] + E_x\left[\int_0^{\tau\wedge t}\left(\frac{1}{2}\lambda(1-v(B_s))-v(B_s)\right)e(B_s)b(B_s)ds\right]. \tag{4.10}$$

Let $\delta = \epsilon\|\phi\|_\infty$. Choose ϵ so that $F'(1-\delta)-1-\frac{\lambda}{2}>0$. Then we have $F(v)\le 1-(1-v)F'(1-\delta)$. Now subtracting (4.10) from (4.9) we obtain

$$\epsilon(t,x)\le F'(1-\delta)E_x\left[\int_0^{\tau\wedge t}\epsilon(t-s,B_s)b(B_s)e(s)ds\right].$$

where $\epsilon(t,x) = \max\{v(t,x)-v(x),0\}$. From this it follows that $\epsilon(t,x)\equiv 0$ (see the end of the proof of Lemma 2.2).

Acknowledgement Thanks are due to Professor Henry McKean for discussions on this problem. The results of this note was presented in a seminar at the Institute For Mathematics and Its Applications at University of Minnesota.

References

[1] Athreya K.B., Ney, P.E., *Branching Processes* , Springer-Verlag, New York, 1972.

[2] Ikeda, N., Nagasawa, M., Watanabe, S., Branching Markov Processes, J. of Math. Kyoto Univ. Vol. 8(1968), 223–278, 365–410, Vol.9, 95–160.

[3] Kenig, Carlos E., Ni, Wei-ming, On the Existence and Boundary Behavior of Solutions to a Class of Nonlinear Dirichlet Problems, Proc. AMS, Vol.89, Number 2(1983), 254–258.

[4] Nagasawa, M., A Probabilistic Approach to Nonlinear Dirichlet Problem, Lecture Notes in Math. no.511, 184–193.

[5] Watanabe, S., On the Branching Process of Brownian Particles with an Absorbing Boundary, J. of Math. Kyoto Univ. vol.4(1965), 385–398.

Pei Hsu
Courant Institute of Mathematical Sciences
251 Mercer Street
New York, NY 10012

REPRESENTATION OF EXCESSIVE MEASURES

by

K. JANSSEN

Introduction

Representations of excessive measures by entrance laws have been obtained by several authors under various regularity assumptions (e.g. Getoor and Sharpe [5], Dynkin [2], Getoor and Glover [4], Fitzsimmons and Maisonneuve [3]). In the following we give a purely analytic proof of such results in complete generality. The proofs follow to some extent the proofs of Getoor and Glover in [4], except that the systematic use of desintegration of measures seems to simplify and to clarify the arguments. As an application we also obtain the result of Getoor and Glover in [4] on the representation of entrance rules in the non-homogenous situation in greater generality, and we deduce the seemingly new result that such entrance rules correspond in a one-to-one way to excessive measures for the space-time semigroup.

I want to thank R.K. Getoor for many discussions on these and related topics.

§ 1 Representation of excessive measures.

Let $(E,\mathcal{E})$ be a U-space, i.e. E is homeomorphic to a universally measurable subset of a compact metric space equipped with its borel field $\mathcal{E}$.
We also denote by $\mathcal{E}$ the set of positive measurable functions on E.

Let $(P_t)_{t>o}$ be a measurable semigroup of kernels on E and define the associated resolvent $(U_\alpha)_{\alpha\geq o}$ by
$U_\alpha f := \int e^{-\alpha t} P_t f dt$ for f in $\mathcal{E}$.

In this chapter we make the following

ASSUMPTIONS.

a) *The kernels* $(P_t)_{t>o}$ *and* $(U_\alpha)_{\alpha>o}$ *are proper kernels.*

b) *The principle of unicity of charges holds*, i.e. if μ,ν are σ-finite measures on E such that for some $\alpha \geq o$ the measures μU_α and νU_α are σ-finite and equal, then $\mu = \nu$.

Assumption b) implies that $(P_t)_{t>o}$ has no branching points, i.e. if μ is a σ-finite measure such that $\mu U_\alpha = \varepsilon_x U_\alpha$ for some $x \in E$ and $\alpha > o$, then $\mu = \varepsilon_x$.
In particular, $U_\alpha 1 > o$ on E holds for all $\alpha \geq o$.

By elementary calculus and Fubini's theorem the following is easily seen:

Let μ be a σ-finite measure on E such that $(m_\alpha)_{\alpha>o} := (\mu U_\alpha)_{\alpha>o}$ is a family of σ-finite measures. Then $(m_\alpha)_{\alpha>o}$ satisfies the resolvent equation, i.e. we have

$$\text{(R)} \quad m_\alpha = m_\beta + (\beta-\alpha)\, m_\alpha U_\beta$$

$$= m_\beta + (\beta-\alpha)\, m_\beta U_\alpha \qquad \text{for } o < \alpha \leq \beta.$$

Moreover there is an integral respresentation

$$\text{(I)} \quad m_\alpha = \int e^{-\alpha t} \eta_t \, dt \qquad (\alpha > o)$$

where $(\eta_t)_{t>o} = (\mu P_t)_{t>o}$ is an *entrance law*, i.e. a family of σ-finite measures satisfying

$$\text{(E)} \qquad \eta_t P_s = \eta_{t+s} \qquad \text{for } s, t > o.$$

Obviously, whenever $(m_\alpha)_{\alpha>o}$ is a family of measures satisfying (R), then $m_o := \sup_{\alpha>o} m_\alpha = \lim_{\alpha\to o} m_\alpha$ is a measure and (R) holds also for $\alpha = o$ under suitable finiteness assumptions. If (I) holds for measures $(m_\alpha)_{\alpha>o}$ and $(\eta_t)_{t>o}$ then this integral representation remains true for $\alpha = o$.

If the measures m_α are σ-finite for $\alpha > o$, then m_o is a countable sum of finite measures but not necessarily σ-finite (take $m_\alpha := \varepsilon_o U_\alpha$ where (U_α) denotes the resolvent of Brownian motion on $\mathbb{R}$).

Remember, that a measure m is called *supermedian* (with respect to (U_α)) if $\alpha m U_\alpha \leq m$ for all $\alpha > o$. m is called *excessive* if $m = \sup_{\alpha>o} \alpha m U_\alpha = \lim_{\alpha\to\infty} \alpha m U_\alpha$.

If $(m_\alpha)_{\alpha>o}$ satisfies (R), then obviously m_{α_o} is supermedian with respect to the resolvent $(U_{\alpha+\alpha_o})_{\alpha>o}$ for each $\alpha_o \geq o$.
In fact, it is well known that under the assumption $U_\alpha 1 > o$ for $\alpha > o$ every supermedian measure is an excessive measure (c.f. 38 in chap. XII in [1]) but we shall not use this result.
We want to prove the following

<u>THEOREM 1</u>. *Let* $(m_\alpha)_{\alpha>o}$ *be a family of* σ*-finite measures such that the resolvent equation* (R) *holds. Then there exists a unique entrance law* $(\eta_t)_{t>o}$ *such that*

$$m_\alpha = \int e^{-\alpha t} \eta_t dt \quad \text{for} \quad \alpha \geq o.$$

Moreover, $\eta_t U_\alpha$ *is* σ*-finite for* $t > o, \alpha > o$.

As a simple application we obtain the following

<u>COROLLARY</u>. *Let* m *be a* σ*-finite excessive measure such that* $m P_t h \to o$ *for* $t \to \infty$ *for some strictly positive* h *in* E. *Then*

$$m = \int \eta_t dt$$

for some entrance law $(\eta_t)_{t>o}$.

Proof of the corollary: Define σ-finite measures $m_\alpha := m - \alpha m U_\alpha$ for $\alpha > o$. The resolvent equation for (U_α) implies that $(m_\alpha)_{\alpha>o}$ satisfies (R), hence the result follows from the above theorem if we can show $m_\alpha \uparrow m$. This is true, since

$$\alpha\ m\ U_\alpha\ h = \alpha \int e^{-\alpha t} m\ P_t h dt = \int e^{-t} m\ P_{t/\alpha}\ hdt$$

decreases to zero by the assumption on h and by dominated convergence.

REMARKS.

1. As we shall see in a moment, the proof of theorem 1 can easily be reduced to the proof of the corollary under the additional assumption that $m\,U_o$ is a probability measure. The essential tool for the proof is a well known result on desintegration of measures and a result on bimeasures. The result as well as the proofs go through under the assumption "E is a separable metric space" instead of "E is a U-space", provided the relevant measures are tight.

2. It is easily seen, that for a σ-finite excessive measure m the assumption $m\ P_t h \to o$ for $t \to o$ for some strictly positive h is equivalent with the existence of a family $(m_\alpha)_{\alpha>o}$ of σ-finite measures satisfying (R) and $\sup_{\alpha>o} m_\alpha = m$.

Theorem 1 associates entrance laws with some excessive measures which are not necessarily σ-finite.

3. The following converse of Theorem 1 can be verified by routine calculus:

If $(\eta_t)_{t>o}$ is an entrance law such that $\eta_t U_\alpha$ is σ-finite for $t,\alpha>o$, then

$$m_\alpha := \int e^{-\alpha t} \eta_t dt \qquad (\alpha>o)$$

defines a family of σ-finite measures satisfying (R).

As one can see from simple examples for uniform motion on $\mathbb{R}$, it is not sufficient to assume only that the measures η_t are σ-finite.

4. The integral formula for m_α in Theorem 1 yields immediately that m_α is an α-excessive measure, since

$$e^{-\alpha s} m_\alpha P_s = \int e^{-\alpha(t+s)} \eta_{t+s} dt = \int 1_{]s,\infty[}(t) e^{-\alpha t} \eta_t dt$$

increases to m_α for s decreasing to zero.

<u>Proof of theorem 1</u>:

1. Step: Reduction of the problem: We may assume $m_\alpha(E)>o$ for $\alpha>o$.

We show that we may assume without loss of generality $U_o 1 < \infty$ and $m_o U_o (E) = 1$, and that it suffices to prove the corollary under this assumption.

Therefore fix any $\alpha > o$ and pick some finite strictly positive h_o in $\mathcal{E}$ such that $U_\alpha h_o < \infty$ and $m_\alpha h_o < \infty$.

Then $m_\alpha U_\alpha h_o \leq m_{\alpha/2} U_\alpha h_o \leq \frac{2}{\alpha} m_\alpha (h_o) < \infty$, i.e. $h := \dfrac{h_o}{m_\alpha U_\alpha h_o}$ is strictly positive and satisfies

$m_\alpha U_\alpha h = 1$, $m_\alpha h < \infty$, $U_\alpha h < \infty$. It is easily verified, that

$$Q_t (f) := \frac{1}{h} e^{-\alpha t} P_t (hf)$$

is a measurable semigroup with resolvent $(V_\beta)_{\beta \geq o}$ given by $V_\beta(f) = \frac{1}{h} U_{\alpha+\beta}(hf)$ for $\beta \geq o$. (V_β) satisfies the principle of unicity of charges, and the family $(n_\beta)_{\beta \geq o}$ defined by $n_\beta := h\, m_{\alpha+\beta}$ is a family of σ-finite measures, which satisfies (R) with respect to the resolvent $(V_\beta)_{\beta \geq o}$, and we have $V_o 1 < \infty$ and $n_o V_o (E) = 1$.

Assume that we can prove the existence and uniqueness of a family $(\sigma_t)_{t>o}$ of σ-finite measures such that $\sigma_t Q_s = \sigma_{t+s}$ for $s,t>o$ and $n_o = \int \sigma_t dt$. Then $\eta_t := \dfrac{e^{\alpha t}}{h} \sigma_t$ defines σ-finite measures such that $\eta_t P_s = \eta_{t+s}$ for $t,s > o$; moreover the family of measures defined by

$$m'_\beta := \int e^{-\beta t} \eta_t \, dt \qquad (\beta > o)$$

satisfies (R) for $(U_\beta)_{\beta>o}$ by routine calculus, and

$$m'_\alpha = \int e^{-\alpha t} \eta_t \, dt = \frac{1}{h} \int \sigma_t \, dt = \frac{1}{h} n_o = m_\alpha .$$

But then we conclude from (R) $m_\beta = m'_\beta$ i.e. $m_\beta = \int e^{-\beta t} \eta_t dt$ for all $\beta > o$.

Moreover we conclude for $t,\beta>o$

$$\eta_t U_\beta = \iint e^{-\beta s}\eta_t \, P_s \, dsdt =$$

$$= \iint e^{-\beta s}\eta_s \, P_t \, dsdt = m_\beta P_t,$$

i.e. $\eta_t U_\beta$ is σ-finite, η_t is uniquely determined by the principle of unicity of charges, and hence the theorem is proved.

<u>2. Step</u>: Proof of the corollary under the assumption $mU(E) = 1$, $U1 < \infty$ for $U := U_o$.
The proof obtains as a consequence of a lemma and a proposition.

In the following $\mathbb{R}_+^*$ denotes the strictly positive reals equipped with it's borel field $B(\mathbb{R}_+^*)$. For f in E and g in $B(\mathbb{R}_+^*)$ we denote by $f \otimes g$ the function defined by $(x,r) \to f(x)g(r)$ for x in E and r in $\mathbb{R}_+^*$.

<u>LEMMA</u>: *There exists a unique measure* ρ *on* $E \times \mathbb{R}_+^*$ *such that*

$$\rho\,(\,A \times \,]t,\infty[\; = m\,P_t 1_A \quad \text{for} \quad t>o,\ A \text{ in } E.$$

Moreover, we have for f *in* E *and* g *in* $B(\mathbb{R}_+^*)$

$$\rho\,(\,Uf\otimes g\,) \quad = \iint g(t)\,P_t f \,dt\,dm;$$

in particular, we have $\rho\,(U1 \otimes 1) = 1$.

<u>Proof</u>: a) For $A \in E$

$$F_A\,(t) := \;m\,P_t(1_A U1) \quad (t>o)$$

defines a finite decreasing right continuous function on $\mathbb{R}_+^*$.

Moreover, $F_A(o+) = m\ (1_A U1) \leq 1$ and $\lim_{t\to\infty} F_A(t) = o$ by dominated convergence. Consequently there exists a subprobability measure

$$\mu_A \text{ on } \mathbb{R}_+^* \text{ such that } \mu_A(\ ]t,\infty[\ ) = F_A(t) \text{ for } t>o.$$

b) The function

$$\rho_o\ (A \times B) := \mu_A(B) \quad \text{for } A \in \mathcal{E},\ B \in \mathcal{B}\ (\mathbb{R}_+^*)$$

defines a bimeasure on $E \times \mathbb{R}_+^*$, i.e.

i) $\rho_o(E \times \mathbb{R}_+^*) = mU1 = 1$;

ii) for $A \in \mathcal{E}$, $B \to \rho_o\ (A \times B) = \mu_A\ (B)$ is a measure on $\mathbb{R}_+^*$;

iii) for $B \in \mathcal{B}(\mathbb{R}_+^*)$, $A \to \rho_o\ (A \times B)$ is a measure on E (this follows by the monotone class theorem from i) and ii), since it is true for sets $B =]t,\infty[$).

From Dellacherie and Meyer (chap. III.T74, p.129 in [1]) we conclude that there exists a unique probability measure ρ_1 on $(E \times \mathbb{R}_+^*,\ \mathcal{E} \otimes \mathcal{B}(\mathbb{R}_+^*))$ such that

$$\rho_1(\ A \times B\ ) = \rho_o\ (A \times B) \quad \text{for } A \in \mathcal{E},\ B \in \mathcal{B}(\mathbb{R}_+^*).$$

Let ρ be the measure on $E \times \mathbb{R}_+^*$ which has density $\frac{1}{U1} \otimes 1$ with respect to ρ_1. Then we have

$$\rho\ (U1 \otimes 1) = \rho_1\ (E \times \mathbb{R}) = 1;\ \text{moreover}$$

$$\rho\ (\ A \times]t,\infty[) = m\ P_t\ 1_A$$

holds for $A \in E$ and $t > o$, and ρ is uniquely determined by these equalities.

Finally, we conclude for f in E and $t > o$

$$\rho\ (Uf \otimes 1_{]t,\infty[}) = m\ P_t Uf = \iint 1_{]t,\infty[}(s)\ P_s f ds dm$$

hence $\rho\ (Uf \otimes g) = \iint g\ (s)\ P_s f\ ds\ d\ m$

for all $g \in B\ (\mathbb{R}_+^*)$ by the monotone class theorem.

<u>PROPOSITION</u>. *There exists a unique family* $(\eta_t)_{t>o}$ *of* σ*-finite measures on* E *such that*

$$\eta_t U = m\ P_t \quad \text{for} \quad t > o.$$

<u>Proof</u>: Let ρ_1 be the probability measure on $E \times \mathbb{R}_+^*$ with density $U1 \otimes 1$ with respect to the measure ρ of the above lemma. Let T denote the projection of $E \times \mathbb{R}_+^*$ onto $\mathbb{R}_+^*$, and denote by μ the image measure of ρ_1 under T. According to Dellacherie and Meyer (chap.III, p. 124-129 in [1])there exists a measurable family $(\sigma_t)_{t>o}$ of probability measures on $E \times \mathbb{R}_+^*$ such that $\rho_1 = \int \sigma_t\ \mu(dt)$ and $\sigma_t\ (\ E \times \{t\}\ ) = 1$ for $t > o$.

According to the above lemma we have for B in $B(\mathbb{R}_+^*)$

$$\mu\ (B) = \rho_1\ (\ T^{-1}(B)\ ) = \rho_1\ (E \times B)$$

$$= \rho(U1 \otimes 1_B) = \iint 1_B(t)\ P_t 1\ dtdm;$$

i.e. μ has density $g_1 : t \to mP_t 1$ with respect to Lebesgue measure on $\mathbb{R}_+^*$ by Fubini's theorem.

Define the measure η'_t on E such that $\eta'_t \otimes \varepsilon_t$ has density $\frac{1}{U1} \otimes g_1$ with respect to σ_t for all $t > o$.

Then we have for $F \in \mathcal{E} \otimes \mathcal{B}(\mathbb{R}_+)$

$$\rho\ (F) = \rho_1\ \left(\frac{F}{U1 \otimes 1} \right) = \iint F\ (x,t)\ \eta'_t\ (dx)\ dt.$$

Applying this formula to the function F_λ defined by $F_\lambda(x,t) := e^{-\lambda t} Uf(x)$ for $x \in E$, $\lambda, t > o$ and $f \in \mathcal{E}$ and using the above lemma gives

$$\int e^{-\lambda t} m\ P_t f dt = \rho\ (F_\lambda) = \iint e^{-\lambda t} Uf(x) \eta'_t\ (dx) dt =$$

$$= \int e^{-\lambda t} \eta'_t Uf\ dt,$$

i.e. $t \to mP_t f$ and $t \to \eta'_t\ Uf$ have the same Laplace-transform, and consequently these two functions are equal almost everywhere with respect to Lebesgue measure if f is bounded by some multiple of $U1$. Letting f run through a suitable countable subset of $\mathcal{E}$ we conclude $mP_t = \eta'_t U$ for almost all $t > o$. In particular we have $mP_{t_n} = \eta'_{t_n} U$ for a suitable sequence $(t_n) \subset \mathbb{R}^*_+$ decreasing to o.

Let $t > o$. For $o < t_k \leq t_n < t$ we have

$$\eta'_{t_k}\ P_{t-t_k}\ U = \eta'_{t_k}\ U\ P_{t-t_k} = m\ P_{t_k}\ P_{t-t_k} =$$

$$= m\ P_t = \eta'_{t_n}\ P_{t-t_n}\ U\ ,$$

hence the measure $\eta_t := \eta'_{t_k}\ P_{t-t_k}$ is well defined, η_t is σ-finite and satisfies $\eta_t U = m\ P_t$, and η_t is uniquely determined by the principle of unicity of charges.

CONSEQUENCES.

a) *We have* $\eta_t\ P_s\ U = \eta_t\ U\ P_s = m\ P_t P_s = m\ P_{t+s} = \eta_{t+s} U$,

hence $\eta_t\ P_s = \eta_{t+s}$ *for* $s,t > o$ *by the principle of unicity of charges.*

b) $$\int \eta_t dt = \lim_{s\to o} \int 1_{]s,\infty[}(t)\ \eta_t(dt) =$$

$$= \lim_{s\to o} \int 1_{]s,\infty[}(t)\ \eta_s\ P_{t-s}\ dt$$

$$= \lim_{s\to o} \int \eta_s\ P_t\ dt = \lim_{s\to o} \eta_s\ U$$

$$= \lim_{s\to o}\ m\ P_s = m,$$

hence the proof of theorem 1 is complete.

§ 2 Entrance rules in the non-homogenous situation

ASSUMPTIONS.

a) We assume that $(E,\mathcal{E})$ is a U-space equipped with its borel field and that $(P_s^r)_{\substack{r,s\in\mathbb{R}\\ r<s}}$ is a family of proper kernels on E satisfying $P_s^r\ P_u^s = P_u^r$ for $r < s < u$.

Moreover, we assume that

$$(x,r,t) \to P_{r+t}^r\ f(x)$$

is $\mathcal{E} \otimes \mathcal{B}(\mathbb{R}) \otimes \mathcal{B}(\mathbb{R}_+^*)$ measurable for f in $\mathcal{E}$, and that for some strictly positive $H \in \mathcal{E} \otimes \mathcal{B}(\mathbb{R})$

$$U\ H\ (x,r) = \int H\ (y,t+r)\ P_{t+r}^r\ (x,dy)\ dt < \infty$$

for all (x,r) in $E \times \mathbb{R}$.

In other words, *we assume that*

$$\varepsilon_{(x,r)} P_t := \varepsilon_x P^r_{r+t} \otimes \varepsilon_{r+t}$$

defines a measurable semigroup $(P_t)_{t>o}$ *of proper kernels on* $E \times \mathbb{R}$ *such that the potential kernel* U *is proper* (a sufficient condition for this is $P^r_s 1 \leq e^{\alpha(s-r)}$ for all $r < s$ for some $\alpha \in \mathbb{R}$).

b) We assume that U *satisfies the principle of unicity of charges*, i.e. if $\mu U = \nu U$ are σ-finite measures for some σ-finite measures μ,ν on $E \times \mathbb{R}$, then $\mu=\nu$.

Remember that a family $(m_s)_{s\in\mathbb{R}}$ of σ-finite measures on $(E,\mathcal{E})$ is called an *entrance rule for* (P^r_s) if for each s in $\mathbb{R}$, $m_r P^r_s$ increases to m_s as r increases to s. A family $(\eta_s)_{s\in\mathbb{R}}$ of σ-finite measures on $(E,\mathcal{E})$ is called an *entrance law at* r in $\mathbb{R}$, if $\eta_s = o$ for $s \leq r$ and $\eta_s P^s_u = \eta_u$ for $r < s < u$.

<u>LEMMA</u>. *Let* (m_s) *be an entrance rule. Then* $s \to m_s f$ *is borel measurable on* $\mathbb{R}$ *for every* f *in* $\mathcal{E}$. *Moreover,*

$$m := \int m_s \otimes \varepsilon_s \, ds$$

is a σ*-finite excessive measure for the space-time semigroup* (P_t).

<u>Proof</u>: Let $s_{n,i} = \frac{i}{2^n}$ for $n\in\mathbb{N}$ and $i\in\mathbb{Z}$, and define for $s_{n,i} < s \leq s_{n,i+1}$

$$\rho^{(n)}_s := m_{s_{n,i}} P^{s_{n,i}}_s .$$

Then $s \to \rho_s^{(n)}(f)$ is measurable for $f \in E$, hence $s \to m_s(f) = \lim \rho_s^{(n)}(f)$ is measurable. Moreover

$$m P_t = \int m_s P_{s+t}^s \otimes \varepsilon_{s+t} \, ds =$$

$$= \int m_{s-t} P_s^{s-t} \otimes \varepsilon_s ds$$

increases to m if t decreases to zero by Beppo-Levi. To prove σ-finiteness of m, pick for each $r \in \emptyset$ a strictly positive function h_r in E such that $\sum_{r \in \emptyset} m_r(h_r) \leq 1$ and define for $s \in \mathbb{R}$ the function

$$f_s := \sum_{\substack{r \in \emptyset \\ s<r}} P_r^s h_r .$$

Then we have

$$m_s f_s \leq \sum_{\substack{s<r \\ r \in \emptyset}} m_r h_r \leq 1 .$$

Consequently $H(x,s) := e^{-|s|} f_s(x)$ defines a measurable function H which is strictly positive, since $H(x,s) = o$ implies $P_r^s h_r(x) = o$ for all r in $\emptyset$, $r > s$, hence $\varepsilon_x P_r^s = o$ for all $s<r$ in $\mathbb{R}$, i.e. $(\varepsilon_x \otimes \varepsilon_s) U = o$ in contradiction with the principle of unicity of charges.

Obviously, we have $m H < \infty$, i.e. m is σ-finite.

<u>THEOREM 2</u>. *Let* m *be a* σ*-finite excessive measure on* $E \times \mathbb{R}$ *for the space-time semigroup* (P_t) *such that* $\lim_{t \to \infty} m P_t H = o$ *for some strictly positive* H *in* $E \otimes B(\mathbb{R})$.

Then there exists a finite measure μ *on* E *and a family* $(\eta_s^r)_{r,s \in \mathbb{R}}$ *of* σ*-finite measures on* E *such that*

i) *for each* $r \in \mathbb{R}$, $(\eta_s^r)_{s \in \mathbb{R}}$ *is an entrance law at* r;

ii) $m_s := \int \eta_s^r \, \mu(dr)$ $(s \in \mathbb{R})$ *defines an entrance rule;*

iii) $m = \int m_s \otimes \varepsilon_s ds$.

Proof: We know that

$$m = \int_0^\infty \eta_t dt,$$

where $(\eta_t)_{t>0}$ is an entrance law for (P_t). The measures η_s^r will be obtained by a regularization of a desintegration of (η_t). We may assume $\eta_t \neq 0$ for some $t > 0$.

a) Desintegration of η_t :

Define for $t \geq 0$ the mapping $T_t : E \times \mathbb{R} \to \mathbb{R}$ by $T_t(x,s) := s - t$.

For $0 < t < u$ we have $\eta_t P_{u-t} = \eta_u$, hence we obtain for $B \in \mathcal{B}(\mathbb{R})$ such that $0 = (T_t(\eta_t))(B) = \eta_t(E \times (B+t))$ the equality

$$(T_u(\eta_u))(B) = \eta_u(E \times (B+u)) = \int 1_B(s-u) d\eta_u(x,s)$$

$$= \int P_{s+u-t}^s \, 1(x) 1_B(s-t) d\eta_t(x,s) = 0,$$

i.e. $T_u(\eta_u)$ is absolutely continuous w.r. to $T_t(\eta_t)$. Let $(t_n) \subset \mathbb{R}_+^*$ decrease to zero, choose $(H_n) \subset \mathcal{E} \otimes \mathcal{B}(\mathbb{R})$ strictly positive such that $\sum_n \eta_{t_n}(H_n) = 1$, and let $\bar{\mu}$ be the probability measure defined by

$$\bar{\mu}(F) := \sum_n \int \eta_{t_n}(H_n F) \quad \text{for } F \in \mathcal{E} \otimes \mathcal{B}(\mathbb{R}).$$

For every $t>0$, $T_t(\eta_t)$ is absolutely continuous with respect to $\mu := T_0(\bar{\mu})$ hence we conclude as in the proof of Theorem 1 from the result on desintegration of measures

applied to $H\eta_t$ for suitable $H>o$ in $\mathcal{E} \otimes B(\mathbb{R})$ and T_t (c.f. chap. III, p. 124-129 in [1]) :
for every $t>o$ there exists a measurable family $(\gamma_{t,r})_{r\in\mathbb{R}}$ of σ-finite measures on E such that

$$\eta_t = \int \gamma_{t,r} \otimes \varepsilon_{t+r} \; \mu(dr);$$

moreover, this family of measures is μ-almost surely uniquely determined.

Since $\eta_{t+u} = \eta_t P_u$, we conclude for fixed $u,t>o$ in $\mathbb{R}$:

$$(*) \qquad \gamma_{t,r} P^{t+r}_{t+r+u} = \gamma_{t+u,r} \quad \text{for } \mu\text{-almost all } r \in \mathbb{R}.$$

Consequently, there exists some $B\in B(\mathbb{R})$ such that $\mu(\mathbb{R} \setminus B) = o$ and

$$\gamma_{t,r} P^{t+r}_{t+r+u} = \gamma_{t+u,r} \quad \text{for } r\in B,\ u,t\in\mathbb{Q}, u,t>o.$$

b) Construction of the entrance laws (η^r_s) :

Let $r\in B$ and $t>o$.

If $o<u<v<t$, $u,v\in\mathbb{Q}$, then

$$\gamma_{u,r} P^{u+r}_{t+r} = \gamma_{u,r} P^{u+r}_{v+r} P^{v+r}_{t+r} = \gamma_{v,r} P^{v+r}_{t+r}$$

i.e. the measure $\delta_{t,r} := \gamma_{u,r} P^{u+r}_{t+r}$ for $o<u<t, u\in\mathbb{Q}$

does not depend on the particular choice of $u<t$ in $\mathbb{Q}$.

If we define

$$\eta^r_s := \begin{cases} \delta_{s-r,r} & \text{if } r\in B,\ s\in\mathbb{R},\ s>r \\ o & \text{otherwise} \end{cases}$$

then $(\eta^r_s)_{s\in\mathbb{R}}$ is an entrance law at $r\in\mathbb{R}$: for $r\in B$

and $s,t \in \mathbb{R}$ such that $s > r$ and $t > o$ we have for every $u \in \mathbb{Q}$ satisfying $o<u<s-r$

$$\eta_s^r \; P_{s+t}^s = \gamma_{u,r} \; P_s^{u+r} \; P_{s+t}^s = \gamma_{u,r} \; P_{s+t}^{u+r} = \eta_{s+t}^r,$$

the other relations for an entrance law beeing obvious.

c) Proof of the result:

We have for (α_n) in $\mathbb{Q}$ decreasing to zero by (*) in a) and by b)

$$m = \int_o^\infty \eta_t dt = \int_o^\infty \int \gamma_{t,r} \otimes \varepsilon_{t+r} \; \mu(dr)dt$$

$$= \sup_n \int\int 1_{]\alpha_n,\infty[}^{(t)} \; \gamma_{\alpha_n,r} \; P_{t+r}^{\alpha_n+r} \otimes \varepsilon_{t+r} \; \mu(dr)dt$$

$$= \sup_n \int\int 1_{]\alpha_n,\infty[}^{(t)} \; \eta_{t+r}^r \otimes \varepsilon_{t+r} \; \mu(dr)dt$$

$$= \int (\int \eta_v^r \; \mu(dr)) \otimes \varepsilon_v \; dv.$$

If we define $m_v := \int \eta_v^r \; \mu(dr)$ for v in $\mathbb{R}$, then we have $m = \int m_v \otimes \varepsilon_v \; dv$; moreover $(m_v)_{v \in \mathbb{R}}$ is an entrance rule, since for $u < v$ in $\mathbb{R}$

$$m_u \; P_v^u = \int \eta_u^r \; P_v^u \; \mu(dr) = \int 1_{]-\infty,u[}^{(r)} \; \eta_v^r \; \mu(dr)$$

increases to $\int 1_{]-\infty,v[}^{(r)} \; \eta_v^r \; \mu(dr) = m_v$ for u increasing to v.

<u>COROLLARY 1.</u> *Let* $(m_s)_{s \in \mathbb{R}}$ *be an entrance rule such that* $\lim_{t\to\infty} m_{s-t} \; P_s^{s-t} = o$ *for* s *in* $\mathbb{R}$. *Then there exists a finite measure* μ *on* $\mathbb{R}$ *and an entrance law* $(\eta_s^r)_{s \in \mathbb{R}}$ *at* r *for each* r *in* $\mathbb{R}$ *such that*

$$m_s = \int \eta_s^r \mu(dr) \quad \text{for every } s \in \mathbb{R} .$$

Proof: By the above lemma $m := \int m_s \otimes \varepsilon_s \, ds$ is a σ-finite excessive measure for (P_t). Moreover,

$$mP_t = \int m_{s-t} P_s^{s-t} \otimes \varepsilon_s \, ds$$

converges to the zero measure by dominated convergence, hence the above theorem 2 gives us a finite measure μ on $\mathbb{R}$ and for each $r \in \mathbb{R}$ an entrance law $(\eta_s^r)_{s \in \mathbb{R}}$ at r such that

$$m = \int (\int \eta_s^r \mu(dr)) \otimes \varepsilon_s ds,$$

and moreover $(\int \eta_s^r \mu d(r))_{s \in \mathbb{R}}$ is an entrance rule.
As a consequence of these two representations for m we conclude

$$m_s = \int \eta_s^r \mu(dr)$$

for Lebesgue almost all s in $\mathbb{R}$, hence for all s in $\mathbb{R}$, since both sides are entrance rules.

REMARK. This result has been proved under stronger regularity assumptions directly (i.e. without going back to the representation of excessive measures in the homogenous situation) by Getoor and Glover in [4].

COROLLARY 2. *There is a one-to-one correspondence between σ-finite excessive measures m for the space-time semigroup on $E \times \mathbb{R}$ and entrance rules $(m_s)_{s \in \mathbb{R}}$ on E. This correspondence is given by*

$$m = \int m_s \otimes \varepsilon_s ds$$

<u>Proof</u>: By the above lemma, this formula defines a σ-finite excessive measure m, if (m_s) is an entrance rule.

Conversely, if m is a σ-finite excessive measure, then $m_o := \lim_{t\to\infty} m P_t$ is a σ-finite excessive measure below m such that $m - m_o$ satisfies the assumptions of theorem 2. Consequently, we may assume without loss of generality $m = m_o$, i.e. $m = m P_t = \alpha m U_\alpha$ for all $t > o$, $\alpha > o$. Let $T : E \times \mathbb{R} \to \mathbb{R}$ be the projection onto $\mathbb{R}$. Then we have for $B \in \mathcal{B}(\mathbb{R})$ and $\alpha > o$

$$(T(m))(B) = m(E \times B) = \alpha m U_\alpha (E \times B) =$$

$$= \alpha \int_o^\infty e^{-\alpha t} m P_t(E \times B)\, dt$$

$$= \alpha \int_o^\infty \int_{E \times \mathbb{R}} e^{-\alpha t} 1_B (s+t) P^s_{s+t} 1(x)\, dm(x,s)\, dt$$

$$= o,$$

if B is of Lebesgue measure zero. From the desintegration theorem we conclude as in the above proofs

$$m = \int m'_s \otimes \varepsilon_s \, ds$$

for a suitable "uniquely determined" measurable family $(m'_s)_{s \in \mathbb{R}}$ of σ-finite measures on E. Moreover we conclude from $m P_\alpha = m$ for each $\alpha > o$

$$(*) \quad m'_{s-\alpha} P^{s-\alpha}_s = m'_s \quad \text{almost everywhere [ds].}$$

Moreover, we have

$$\int m'_s \otimes \varepsilon_s ds = m = m U_1 =$$

$$= \int\int_0^\infty e^{-t} m'_s P^s_{s+t} \otimes \varepsilon_{s+t} \, dtds$$

$$= \int_0^\infty\int e^{-t} m'_{s-t} P^{s-t}_s \otimes \varepsilon_s \, dsdt$$

$$= \int m_s \otimes \varepsilon_s \, ds$$

$$\text{for } m_s := \int_0^\infty e^{-t} m'_{s-t} P^{s-t}_s \, dt .$$

Finally, (m_s) is an entrance law, since (*) implies for every s in $\mathbb{R}$ and $\alpha>o$

$$m_s = \int_0^\infty e^{-t} m'_{s-t} P^{s-t}_s \, dt$$

$$= \int_0^\infty e^{-t} m'_{s-t-\alpha} P^{s-t-\alpha}_{s-t} P^{s-t}_s \, dt$$

$$= \int_0^\infty e^{-t} m'_{s-\alpha-t} P^{s-\alpha-t}_{s-\alpha} P^{s-\alpha}_s \, dt$$

$$= m_{s-\alpha} P^{s-\alpha}_s$$

REFERENCES.

1. C.Dellacherie, P.A. Meyer: Probabilités et potentiel. Chap. I à IV, Hermann, Paris, 1975.
   Chap. XII-XIII, preprint.

2. E.B. Dynkin. Minimal excessive measures and functions. Trans.Amer. Math. Soc. 258 (1980) 217-244.

3. P.J. Fitzsimmons and B. Maisonneuve. Excessive measures and Markov processes with random birth and death. Probab.Th. Rel. Fields 72 (1986) 319-336.

4. R.K. Getoor, J. Glover: Constructing Markov processes with random times of birth and death. To appear in: Seminar on Stochastic Processes 1986. Birkhäuser. Boston-Basel-Stuttgart.

5. R.K. Getoor and M.J. Sharpe. Last exit times and additive functionals. Ann. Prob. 1 (1973) 550-569.

Klaus Janßen
Institut für Statistik und Dokumentation

Universitätsstraße 1

UNIVERSITÄT DÜSSELDORF

4000 Düsseldorf-1

THE EXACT HAUSDORFF MEASURE OF BROWNIAN MULTIPLE POINTS

by

J.F. LE GALL

1. INTRODUCTION.

Let $B = (B_t, t \geq 0)$ denote a d-dimensional Brownian motion, with $d \geq 2$. Dvoretzky, Erdös, Kakutani and Taylor [2,3,4] have proved that the path of B has points of multiplicity $p \geq 2$ if and only if one of the two following conditions is satisfied :

(1.a)
- $d = 2$, p arbitrary
- $d = 3$, $p = 2$.

A natural problem is then to evaluate the size of the set D_p of points of multiplicity p. The Hausdorff dimension of D_p was obtained by Taylor [20], for $d = 2$, and Fristedt [7], for $d = 3$:

- if $d = 2$, $\dim(D_p) = 2$, for any p
- if $d = 3$, $\dim(D_2) = 1$.

Note that, in the case $d = 2$, the dimension of D_p is the same for all integers p's , while it is intuitively clear that there are much more points of multiplicity p than points of multiplicity $p + 1$.

This intuitive statement can be made rigorous by considering the Hausdorff measure of D_p with respect to some suitable measure function. For any $r \in \mathbb{R}$, let $g_r(x) = x^2(\log 1/x)^r$. Then, still assuming that $d = 2$,

$$(1.b) \quad g_r - m(D_p) = \begin{cases} 0 & \text{if } r \leq p, \\ \infty & \text{if } r > p, \end{cases}$$

where $g_r - m$ denotes the Hausdorff measure associated with g_r. The latter result was conjectured by Taylor [21] and proved in [9]. The techniques of [9] also yield the following result in the case $d = 3$: if $g_r'(x) = x(\log 1/x)^r$,

$$(1.c) \quad g_r' - m(D_2) = \begin{cases} 0 & \text{if } r \leq 0, \\ \infty & \text{if } r > 0. \end{cases}$$

Our goal here is to improve on the results $(1.b)$ and $(1.c)$ by determining a "correct" measure function for D_p. We will find a function g such that, on one hand, the g-measure of D_p is positive and, on the other hand, D_p is a countable union of sets of finite g-measure. The g-measure of D_p itself will not be finite, unless we restrict ourselves to a particular set of multiple points.

Let us emphasize the fact that there may exist many correct measure functions for a given set. Nevertheless we will refer to *the* correct measure function, as it were unique.

In the case $p = 1$, D_1 is simply the path of B. The Hausdorff measure of D_1 was investigated by several authors, including Lévy [14] and Ray [15]. The correct measure function for D_1 was shown by Ciesielski - Taylor [1] $(d \geq 3)$ and Taylor [19] $(d = 2)$ to be :

- if $d = 2$, $h_1(x) = x^2 \log 1/x \log\log\log 1/x$
- if $d \geq 3$, $k_1(x) = x^2 \log\log 1/x$.

Before we state our main results, we need to introduce the notion of intersection local time, which plays a key role throughout this work. For any $p \geq 2$, let $\mathcal{C}_p$ denote the simplex :

$$\mathcal{C}_p = \{(t_1,\dots,t_p) \in (\mathbb{R}_+)^p \ ; \ 0 \le t_1 < t_2 < \dots < t_p\}.$$

Assume that the pair (d,p) satisfies *(1.a)*. The intersection local time, at the order p, of B with itself, is the Radon measure on $\mathcal{C}_p$ which is formally defined by :

$$\alpha_p(A) = \int_A \delta_{(0)}(B_{t_1}-B_{t_2})\dots\delta_{(0)}(B_{t_{p-1}}-B_{t_p})dt_1\dots dt_p,$$

for any Borel subset A of $\mathcal{C}_p$. Here $\delta_{(0)}$ denotes the Dirac measure at 0. A rigorous definition of α_p can be found in Dynkin [6] or Rosen [17,18]. Note that it is essential that we exclude the diagonal $\{t_1 = \dots = t_p\}$ in the definition of $\mathcal{C}_p$. In fact, elementary scaling arguments suffice to prove that $\alpha_p(\mathcal{C}_p \cap [0;1]^p) = \infty$, a.s. (see [11] for the special case $d = 2$, $p = 2$). As is suggested by the above formal definition, the measure α_p is supported by the p-uples $(t_1,\dots,t_p)$ such that $B_{t_1} = \dots = B_{t_p}$. Let ℓ_p denote the image measure of α_p by the mapping $(t_1,\dots,t_p) \to B_{t_1}$. It follows that ℓ_p is supported by D_p. In some sense, ℓ_p is the canonical measure supported by D_p (see [9]). Simple arguments imply that ℓ_p is non-zero and σ-finite.

With these definitions at hand we are now ready to state our main theorems.

<u>*Theorem 1*</u>. *Assume that $d = 2$, $p \ge 2$. Let h_p denote the function :*

$$h_p(x) = x^2(\log 1/x \log\log\log 1/x)^p.$$

There exist two positive constants C_p, C'_p such that, a.s. for any Borel subset F of $\mathbb{R}^2$,

$$C_p\, \ell_p(F) \le h_p - m(F \cap D_p) \le C'_p\, \ell_p(F).$$

Theorem 2. Assume that $d = 3$. *Let* k_2 *be the function :*

$$k_2(x) = x(\log\log 1/x)^2.$$

There exist two positive constants C, C' *such that, a.s. for any Borel subset* F *of* $\mathbb{R}^3$,

$$C_2\, \ell_2(F) \leq k_2 - m(F \cap D_2) \leq C_2'\, \ell_2(F).$$

If is worth noting that *theorem 1* also holds for $p = 1$. In this case, ℓ_1 should be interpreted as the occupation measure of the Brownian path, which is the image of Lebesgue measure on $\mathbb{R}_+$ by the mapping $t \to B_t$. A similar remark holds for *theorem 2*.

Let us mention an interesting consequence of *theorems 1* and *2* . Let $x^d\, f(x)$ denote the correctHausdorff measure function for the Brownian sample path in $\mathbb{R}^d$.Then the correct measure function for the set of p-multiple points is simply $x^d\, (f(x))^p$. One may ask whether this property could have been established directly, and also whether it can be extended to other processes.

Theorems 1 and *2* do not provide much information for subsets F such that $\ell_p(F) = \infty$. We will now describe some special subsets of D_p whose ℓ_p-measure is positive and finite. For any $\varepsilon, N > 0$ set :

$$\mathcal{C}_p^{\varepsilon,N} = \{(t_1,\dots,t_p) \in \mathcal{C}_p \cap [0;N]^p \ ;\ t_i - t_{i-1} \geq \varepsilon, \text{ for any } i = 2,\dots p\}$$

$$D_p^{\varepsilon,N} = \{y \in \mathbb{R}^2 \ ;\ y = B_{t_1} = \dots = B_{t_p} \text{ for some } (t_1,\dots,t_p) \in \mathcal{C}_p^{\varepsilon,N}\}.$$

It follows from *theorems 1* and *2* that, a.s. for ε small enough (or for N large enough),

(1.d)
- if $d = 2$, $0 < h_p - m(D_p^{\varepsilon,N}) < \infty$,
- if $d = 3$, $0 < k_2 - m(D_2^{\varepsilon,N}) < \infty$.

In order to deduce *(1.d)*, let ϕ denote the mapping $(s_1,\dots,s_p) \to B_{s_1}$, so that :

$$\ell_p(D_p^{\varepsilon,N}) = \alpha_p(\phi^{-1}(D_p^{\varepsilon,N})).$$

It is obvious from the definition that :

$$\mathcal{E}_p^{\varepsilon,N} \subset \phi^{-1}(D_p^{\varepsilon,N}).$$

On the other hand, it was noticed in [13] (see the remark after *corollary 2.2*) that :

(1.e) for α_p-a.a. $(s_1,\dots s_p)$, $B_s \neq B_{s_1}$ if $s \notin \{s_1,\dots,s_p\}$.

In other words, ℓ_p-almost every p-multiple point is exactly p-multiple. It follows that

$$\ell_p(D_p^{\varepsilon,N}) = \alpha_p(\mathcal{E}_p^{\varepsilon,N}) < \infty,$$

since α_p is a Radon measure. On the other hand a zero-one law argument implies that, a.s. for ε small enough, $\alpha_p(\mathcal{E}_p^{\varepsilon,N}) > 0$. The desired result follows.

We now note that : $D_p = \bigcup_{\varepsilon>0,N>0} D_p^{\varepsilon,N}$.

The bounds *(1.d)* demonstrate that, in some sense, h_p, resp. k_2 in the case $d = 3$, is the correct measure function for D_p.

It would be very interesting to replace the statement of *theorem 1* by :

(1.f) $h_p - m(F \cap D_p) = C_p\, \ell_p(F)$,

for some constant C_p and any Borel subset F. Such a result would provide an intrinsic construction of the intersection local time from the set of multiple points. Note that *(1.f)* holds in the case $p = 1$ (see [1], [19]).

Statements equivalent to *theorems 1* and *2* can be obtained by studying the set of intersection points of p independent Brownian paths in $\mathbb{R}^d$. We need to introduce the intersection local time of p independent Brownian motions $B^1,\dots,B^p$, which is the Radon measure on $(\mathbb{R}_+)^p$ formally defined by :

$$\beta_p(A) = \int_A \delta_{(0)}(B^1_{t_1}-B^2_{t_2})\dots\delta_{(0)}(B^{p-1}_{t_{p-1}}-B^p_{t_p})dt_1\dots dt_p,$$

(see Dynkin [5] or Geman - Horowitz - Rosen [8] for a precise definition). Let λ_p denote the image measure of β_p by the mapping $(t_1\dots t_p) \to B^1_{t_1}$, and let I_p be the set of intersection points of the paths of $B^1,\dots,B^p$. *Theorem 1* for instance can be extended as follows: for any Borel subset F of $\mathbb{R}^2$,

$$C_p\ \lambda_p(F) \leq h_p - m(F \cap I_p) \leq C'_p\ \lambda_p(F).$$

Although we will only prove *theorems 1* and *2*, our arguments apply as easily to the situation of intersection points of independent processes. As a matter of fact, in the course of the proofs of *theorems 1* and *2*, we will have to turn statements about multiple points of a single process into equivalent statements concerning intersection points of independent processes. This transformation will be achieved through the results of [13], which provide some information on the behaviour of a Brownian motion between the successive hitting times of a multiple point.

The proofs of *theorems 1* and *2* depend on the density theorems for Hausdorff measures which were established by Rogers and Taylor [16] ; they are recalled at the beginning of section 3. In section 2, we state and prove a few properties of intersection local times to be used in the sequel. In contrast with [9], we do not use the approximation of the intersection local time through Wiener sausages, except for deriving an expression of the moments of intersection local times. Section 3 deals with the case $d = 3$, $p = 2$ which is easier. The case $d = 2$ is developed in section 4 and requires a few auxiliary results on the local behaviour of the intersection local time. Our methods are very close to those which were used in [10] for the case $p = 1$. The latter paper was itself largely inspired by Taylor [19]. The special arguments in the case $d = 2$ owe a lot to Ray's work [15].

The results of this paper can obviously be extended to multiple points of more general Lévy processes. Details will be developed in a forthcoming paper.

Acknowledgments. This work was accomplished in part while the author was visiting the University of Virginia. The author wishes to thank Professors S.J. Taylor and L. Pitt for their hospitality.

2. *PRELIMINARY RESULTS.*

(2.1) Our goal in this section is to establish a few results about intersection local times, which will be needed in the proofs of *theorems 1* and *2*. We consider a pair (d,p) satisfying *(1.a)*. Let $B^1,\dots,B^p$ be p independent Brownian motions with values in $\mathbb{R}^d$. As in the above introduction, we denote by β_p the intersection local time of $B^1,\dots,B^p$, defined by the formal expression *(1.g)*, and by λ_p the associated measure on $\mathbb{R}^d$. Some formal manipulations lead to the expression :

$$(2.a) \qquad \lambda_p(F) = \int_F dy \prod_{i=1}^{p} \left(\int_0^\infty ds\ \delta_{(y)}(B_s^i)\right).$$

We will need the following "localized" version of *(2.a)*. Let J be a Borel subset of $(\mathbb{R}_+)^p$. We denote by β_p^J the measure induced by β_p on J and by λ_p^J the image of β_p^J by the mapping $(t_1,\dots,t_p) \to B_{t_1}$. If we restrict ourselves to the special case when :

$$J = [0;a_1] \times [0;a_2] \times \dots \times [0;a_p]$$

we can replace *(2.a)* by :

$$(2.a') \qquad \lambda_p^J(F) = \int_F dy \prod_{i=1}^{p} \left(\int_0^{a_i} ds\ \delta_{(y)}(B_s^i)\right).$$

Formula *(2.a')* makes it possible to give an explicit expression of the moments of $\lambda_p^J(F)$. Let us emphasize the formal character of formulas *(2.a)*, *(2.a')*, so that the results which will be derived from these formulas must be justified rigorously, using for instance the approximation of λ_p through Wiener sausages. However formulas *(2.a)*, *(2.a')* provide some heuristic explanation for the following results. Specializing to the case when $J = [0;1]^p$ and $B_0^1 = B_0^2 = \dots = B_0^p = 0$, we have (see [12] *proposition 2.1*) : for any $k \geq 1$,

(2.b) $E[(\lambda_p^J(F))^k] =$

$$= \int_{F^k} dy_1\ldots dy_k \Big(\sum_{\sigma\in S_k} \int_{\{0<s_1<\ldots<s_k<1\}} ds_1\ldots ds_k \prod_{i=1}^{k} p_{s_i-s_{i-1}}(y_{\sigma(i-1)},y_{\sigma(i)}) \Big)^p$$

where S_k denotes the set of permutations of $\{1,\ldots,k\}$, $p_s(y,z)$ is the d-dimensional Gaussian transition density, and we agree that $s_o = 0$, $y_{\sigma(0)} = 0$.

We will need an analogue of (2.b) when J is some random subset of $(\mathbb{R}_+)^p$. Note that the definition of λ_p^J makes sense even in this case. Our goal is to establish the following proposition.

Proposition 3. Assume that $|B_0^i| \leq 1$ *for each* i*, and let* $R \geq 2$*. For any* $i \in \{1,\ldots,p\}$*, set*

$$T_R^i = \inf\{t \geq 0 \ ; \ |B_t^i| = R\},$$

and $J = J_R = \prod_{i=1}^{p} [0;T_R^i]$.

Then there exist two positive constants C,C'*, depending on the pair* (d,p) *but not on* R*, such that, for any integer* $k \geq 1$*,*

- *if* $d = 2$*,*

$$C^k(\log R)^{pk}(k!)^p \leq E[\lambda_p^J(D)^k] \leq C'^k(\log R)^{pk}(k!)^p$$

- *if* $d = 3$*,*

$$C^k(k!)^2 \leq E[\lambda_2^J(D)^k] \leq C'^k(k!)^2.$$

Here $D = D(0,1)$ *denotes the closed ball of radius* 1 *centered at* 0 *in* $\mathbb{R}^d$.

Proof : We will only treat the case $d = 2$. The case $d = 3$ is similar and easier. Without loss of generality we may assume $B_o^i = x_i \in D$, for each $i = 1,\ldots,p$. Let $p_s^*(y,z)$ denote the transition density of Brownian motion killed when it hits the complement of the disk $D(0,R)$, and set :

$$G(y,z) = \int_0^\infty p_s^*(y,z)ds \qquad (y,z \in \mathbb{R}^2).$$

The following analogue of (2.b) holds :

(2.c) $E[(\lambda_p^J(D))^k] =$

$$= \int_{D^k} dy_1 \ldots dy_k \prod_{i=1}^{p} \Big(\sum_{\sigma \in S_k} G(x_i, y_{\sigma(1)}) \prod_{j=2}^{k} G(y_{\sigma(i-1)}, y_{\sigma(i)}) \Big).$$

Formula (2.c) can be deduced from (2.a') by some formal considerations. However we will now sketch a rigorous proof of (2.c). Let us introduce the Wiener sausage of radius ε associated with each Brownian motion on the time interval $[0;T_R^i]$:

$$S_\varepsilon^i = \{y \in \mathbb{R}^2 \ ; \ \inf(|B_s^i - y| \ ; \ s \leq T_R^i) \leq \varepsilon\}.$$

The arguments of [9] show that :

$$\lim_{\varepsilon \to 0} (\log \tfrac{1}{\varepsilon}) \, P[y \in S_\varepsilon^i] = \pi \, G(x_i, y).$$

More generally, let $y^1, \ldots, y^k$ be k distinct points of $D(0;R)$, then the strong Markov property yields that ;

$$\lim_{\varepsilon \to 0} (\log \tfrac{1}{\varepsilon})^k \, P[\bigcap_{j=1}^{k} \{y_j \in S_\varepsilon^i\}]$$

$$= \pi^k \sum_{\sigma \in S_k} (G(x_i, y_{\sigma(1)}) \prod_{j=2}^{k} G(y_{\sigma(j-1)}, y_{\sigma(j)})).$$

On the other hand, note that :

$$E[m(S_\varepsilon^1 \cap \ldots \cap S_\varepsilon^p \cap D)^k] = \int_{D^k} dy_1 \ldots dy_k \prod_{i=1}^{p} P[\bigcap_{j=1}^{k} \{y_j \in S_\varepsilon^i\}],$$

and it follows from the results in [9] that :

$$\lim_{\varepsilon \to 0} (\log \tfrac{1}{\varepsilon})^p \, m(S_\varepsilon^1 \cap \ldots \cap S_\varepsilon^p \cap D) = \pi^p \lambda_p^J(D),$$

where convergence holds in L^k-norm for any $k < \infty$.
Putting together the previous results and applying Lebesgue dominated convergence theorem, we get (2.c).

The statement of the proposition will now follow easily from (2.c). We first note that there exist two positive constants c_1, c_1', not depending on R, such that, for any $y, z \in D$,

$$c_1 \log\left(\frac{2R}{|y-z|}\right) \leq G(y,z) \leq c_1' \log\left(\frac{2R}{|y-z|}\right).$$

The lower bound of the proposition follows at once from (2.c) and the above bound for $G(y,z)$. In order to deduce the upper bound, we use Hölder's inequality which implies that :

$$E[(\lambda_p^J(D))^k] \leq (c_1')^{pk} (\log R)^{pk} (k!)^p$$
$$\times \left(\sup_{y\in D} \int_D dz \left(1 + \frac{\log 2/|y-z|}{\log R}\right)^p\right)^k$$
$$\leq C'^k (\log R)^{pk} (k!)^p,$$

where C' is some constant not depending on R. □

We will use the bounds of *proposition* 3 in connection with the following elementary lemma.

Lemma 4. *Let* X *be a nonnegative random variable, and* $\alpha > 0$. *Assume that there exist two positive constants* C_1, C_2 *such that, for any integer* $k \geq 1$,

$$C_1^k k^{\alpha k} \leq E[X^k] \leq C_2^k k^{\alpha k}.$$

Then there exist two positive constants λ_1, λ_2, *depending only on* α, C_1, C_2, *such that, for any* $a > 0$ *large enough,*

$$\exp(-\lambda_1 a^{1/\alpha}) \leq P[X > a] \leq \exp(-\lambda_2 a^{1/\alpha}).$$

Proof : The upper bound is easy. It suffices to notice that, for $\lambda > 0$ small enough,

$$E[\exp(\lambda X^{1/\alpha})] < \infty.$$

In order to get the lower bound, we first note that, possibly with different values of C_1, C_2, we also have :

$$C_1^r r^{\alpha r} \leq E[X^r] \leq C_2^r r^{\alpha r}, \tag{2.d}$$

for any real number $r \geq 1$. Then, using Cauchy - Schwarz inequality,

$$E[X^r] \leq a^{-1}\, E[X^{r+1}\, 1_{(X>a)}] + a^r$$
$$\leq a^{-1}\, E[X^{2(r+1)}]^{1/2}\, P[X > a]^{1/2} + a^r.$$

Hence,

$$P[X > a]^{1/2} \geq a\Big(\frac{E[X^r] - a^r}{E[X^{2(r+1)}]^{1/2}}\Big).$$

We apply the latter bound with $r = \lambda\, a^{1/\alpha}$, where λ is taken so large that $C_1\, \lambda^\alpha > 2$. Then, for any a large enough,

$$P[X > a]^{1/2} \geq a\Big(\frac{C_1^r\, \lambda^{\alpha r}\, a^r - a^r}{C_2^{r+1}(2(r+1))^{\alpha(r+1)}}\Big)$$
$$\geq c_2\, c_3^{-r},$$

for some positive constants c_2, c_3. The desired result follows. □

(2.2) In subsection *(2.1)* we obtained a few results concerning the intersection local time of p independent Brownian motions. We now propose to show how these results can be applied to the intersection local time, at the order p, of one Brownian motion with itself. First of all we will point out that *theorems 1* and *2* may be replaced by equivalent statements involving the values of B on a finite number of disjoint intervals.

We consider a pair (d,p) which satisfies condition *(1.a)*. Let B denote a Brownian motion in $\mathbb{R}^d$ and let α_p, ℓ_p be defined as in section 1. Let J be the subset of $\mathcal{E}_p$ defined by :

(2.e) $\quad J =]a_1 ; a_1'[\times]a_2 ; a_2'[\times \ldots \times]a_p ; a_p'[,$

where $0 < a_1 < a_1' < a_2 < a_2' < \ldots < a_p < a_p'$. Let α_p^J denote the measure induced by α_p on J, and let ℓ_p^J be the image of α_p^J by the mapping $(t_1,\ldots,t_p) \to B_{t_1}$. Set :

$$D_p^J = \{y \in \mathbb{R}^d \; ; \; y = B_{t_1} = \ldots = B_{t_p} \text{ for some } (t_1,\ldots,t_p) \in J\}.$$

Then *theorem 1* for instance may be replaced by the following equivalent statement : there exist two positive constants C, C' such that for any subset J of $\mathcal{C}_p$ of the type (2.e), a.s. for any Borel subset F of $\mathbb{R}^2$,

$$(2.f) \quad C\, \ell_p^J(F) \leq h_p - m(F \cap D_p^J) \leq C'\, \ell_p^J(F).$$

Indeed suppose that (2.f) holds for any subset J.
Then we may find a countable family $(J_m\ ;\ m = 1,2,\ldots)$ of disjoint subsets of $\mathcal{C}_p$, of the type (2.e), such that $\mathcal{C}_p - \bigcup_m J_m$ is contained in a countable union of hyperplanes (a subset of $\mathcal{C}_p$ of the type

$$H = \{(t_1,\ldots t_p),\ t_i = a\},$$

for some $i \in \{1,\ldots,p\}$, $a \geq 0$, is called a hyperplane). Since the α_p-measure of any hyperplane is zero, we have, for any Borel subset F of $\mathbb{R}^2$,

$$(2.g) \quad \ell_p(F) = \sum_m \ell_p^{J_m}(F).$$

On the other hand, it follows from (1.b) that, if $m \neq m'$,

$$h_p - m(D_p^{J_m} \cap D_p^{J_{m'}}) = 0,$$

(this is also a consequence of (1.e) and (2.f)).
Therefore,

$$(2.h) \quad h_p - m(F \cap D_p) = \sum_m h_p - m(F \cap D_p^{J_m}).$$

Using relations (2.g) and (2.h) we see that the statement of *theorem 1* is a consequence of (2.f) . Similar remarks apply to *theorem 2*.

From now on we consider a fixed subset J of the type (2.e) . Set :

$$\bar{J} = [0\ ;\ a_1' - a_1] \times \ldots \times [0\ ;\ a_p' - a_p].$$

Let $X^1,\ldots,X^p$ be p independent Brownian motions in $\mathbb{R}^d$ such that, for each i, the law of X_0^i has a positive density with respect to

Lebesgue measure in $\mathbb{R}^d$. Note that the distribution of :

$$(B(a_1 + s_1), B(a_2 + s_2), \dots, B(a_p + s_p) \; ; \; (s_1, \dots, s_p) \in \bar{J})$$

is then absolutely continuous with respect to that of :

$$(X^1(s_1), X^2(s_2), \dots, X^p(s_p) \; ; \; (s_1, \dots, s_p) \in \bar{J}),$$

and conversely. In particular, we may take advantage of this property to define the intersection local time $\beta_p^{\bar{J}}$ of the p processes $(B(a_i + s_i) \; ; \; 0 \leq s_i \leq a_i' - a_i)$. Let $\lambda_p^{\bar{J}}$ denote the image measure of $\beta_p^{\bar{J}}$ by the mapping $(s_1, \dots, s_p) \to B(a_1 + s_1)$. It should now be clear for instance from the density of occupation time formulas (see [8] and[17,18]), that $\beta_p^{\bar{J}}$ and α_p^J are linked by the following relations:

$$\int f(s_1, \dots, s_p) \; \alpha_p^J(ds_1 \dots ds_p) = \int f(a_1 + s_1, \dots, a_p + s_p) \beta_p^{\bar{J}}(ds_1 \dots ds_p),$$

so that $\lambda_p^{\bar{J}} = \ell_p^J$. This construction establishes the relationship between the intersection local time at the order p, of one process with itself, and the intersection local time of p independent processes . From now on, we will make this identification without further comment.

(2.3) Many difficulties in the study of Brownian multiple points arise from the fact that the Markov property cannot be applied at the first hitting time of a multiple point. As a matter of fact, the first hitting time of a multiple point is never a stopping time. The following proposition, which follows from *corollary 2.2* in [13], provides a way to avoid these difficulties. We need to introduce one more notation. If $X = (X(t) \; ; \; t \geq 0)$ is any process, we set, for $0 \leq u < v$

$${}_uX_v(t) = \begin{cases} X(u+t) - X(u) & \text{if } t \leq v - u \\ X(v) - X(u) & \text{if } t > v - u \end{cases}$$

and

$${}_vX_u(t) = \begin{cases} X(v - t) - X(u) & \text{if } t \leq v - u \\ X(u) - X(v) & \text{if } t > v - u. \end{cases}$$

We consider a set J of the type *(2.e)*.

Proposition 5. *Let $C(\mathbb{R}_+,\mathbb{R}^d)$ denote the space of all continuous functions from $\mathbb{R}_+$ to $\mathbb{R}^d$, and let $X^1,\ldots,X^{2p}$ be $2p$ independent Brownian motions started at 0. Let m denote Lebesgue measure on $(\mathbb{R}_+)^p$ and let H be a Borel subset of $C(\mathbb{R}_+,\mathbb{R}^d)$ such that for m-a.a. $(s_1,\ldots,s_p) \in J$,*

$$({}_0X^1_{s_1-a_1},{}_0X^2_{a_1'-s_1},\ldots,{}_0X^{2p-1}_{s_p-a_p},{}_0X^{2p}_{a_p'-s_p}) \in H, \quad a.s.$$

Then, a.s. for α_p^J-a.a. $(s_1,\ldots,s_p)$,

$$({}_{s_1}B_{a_1},{}_{s_1}B_{a_1'},\ldots,{}_{s_p}B_{a_p},{}_{s_p}B_{a_p'}) \in H.$$

Proposition 5 will be useful whenever we have to verify that some statement holds for ℓ_p^J-almost all multiple points. It will suffice to prove that some property is satisfied a.s. by 2p independent Brownian motions starting from the same point.

3. PROOF OF THEOREM 2.

(3.1) Throughout this section, $B = (B_t\ ;\ t \geq 0)$ denotes a Brownian motion in $\mathbb{R}^3$. We use the notations of section 1 and subsection 2.2. In particular J is a subset of $\mathcal{C}_2$ of the type

(3.a) $\quad J =]a\ ;\ b[\ \times\ ]c\ ;\ d[,$

where $0 < a < b < c < d$. The measures α_2^J, ℓ_2^J, and the set D_2^J were defined in subsection 2.2. It was noticed above that the proof of *theorem 2* reduces to the existence of positive constants C, C', not depending on J, such that, a.s. for any Borel subset F of $\mathbb{R}^3$,

(3.b) $\quad C\ \ell_2^J(F) \leq k_2 - m(F \cap D_2^J) \leq C'\ \ell_2^J(F).$

We now recall the density theorems for Hausdorff measures which were established by Rogers and Taylor [16]. Our formulation is somewhat different from that given in [16]. However it is not hard to see that the arguments of [16] also yield the following result (1).

(1) This extension was pointed out to the author by E. Perkins.

Proposition 6 : Let $h : \mathbb{R}_+ \to \mathbb{R}_+$ be a monotically increasing function. Assume that the ratio $h(2x)/h(x)$ is bounded by $K > 0$. Let μ be a Radon measure on $\mathbb{R}^d$. For any $\lambda > 0$, set :

$$E(\lambda) = \{y \in \mathbb{R}^d \; ; \; \limsup_{a \to 0} \left(\frac{\mu(D(y,a))}{h(a)}\right) \leq \lambda\}$$

There exist two positive constants γ, γ', depending only on d and K, such that for any Borel subset F of $\mathbb{R}^d$,

- *if $F \subset E(\lambda)$,*

$$h - m(F) \geq \gamma \, \lambda^{-1} \, \mu(F),$$

- *if $F \cap E(\lambda) = \emptyset$,*

$$h - m(F) \leq \gamma' \, \lambda^{-1} \, \mu(F).$$

We will apply *proposition 6* with $\mu = \ell_2^J$ and $h = k_2$. In subsections 3.2 and 3.3 we will show the existence of two positive constants C_1, C_1', not depending on J, such that :

$$(3.c) \qquad \ell_2^J(\{y \in \mathbb{R}^3 \; ; \; \limsup_{a \to 0} \left(\frac{\ell_2^J(D(y,a))}{k_2(a)}\right) \geq C_1\}) = 0,$$

$$(3.d) \qquad k_2 - m(\{y \in D_2^J, \; \limsup_{a \to 0} \left(\frac{\ell_2^J(D(y,a)}{k_2(a)}\right) \leq C_1'\}) = 0.$$

The bounds (3.*b*) follow easily by combining (3.c) and (3.*d*) with *proposition 6*. Thus it only remains to prove (3.c) and (3.*d*). In the sequel we will fix a subset J of the type (3.*a*), it will be clear that our constants do not depend on the choice of J.

(3.2) Proof of (3.c). We aim to verify that some property is satisfied by ℓ_2^J-almost all double points. According to *proposition 5* it suffices to prove that the following statement holds almost surely : if B^1, B^2 are two independent Brownian motions in $\mathbb{R}^3$, starting at 0, and if λ_2 denotes the image measure on $\mathbb{R}^3$ of the intersection local time of B^1 and B^2, there exists a constant C_2 such that : a.s.,

$$(3.e) \qquad \limsup_{a \to 0} \left(\frac{\lambda_2(D(0,a))}{k_2(a)}\right) \leq C_2.$$

For any integer $n \geq 1$, set $a_n = 2^{-n}$. It is enough to prove that, a.s.,

$$(3.f) \qquad \limsup_{n \to \infty} \left(\frac{\lambda_2(D(0,a_n))}{k_2(a_n)}\right) \leq C_2.$$

A scaling argument shows that :

$$\lambda_2(D(0,a_n)) \overset{(d)}{=} a_n \, \lambda_2(D(0,1)),$$

hence, for any $c > 0$,

$$(3.g) \qquad P[\lambda_2(D(0,a_n)) \geq c\, k_2(a_n)] = P[\lambda_2(D(0,1)) \geq c(\log n + \log\log 2)^2].$$

Using both *proposition 3* (with $R = \infty$) and *lemma 4*, we may find a positive constant ρ such that, for any $u > 0$ large enough ,

$$(3.h) \qquad P[\lambda_2(D(0,1)) \geq u] \leq \exp(-\rho\, u^{1/2}).$$

Taking c so large that $\rho\, c^{1/2} > 1$, (3.*g*) and (3.*h*) imply that :

$$\sum_{n=1}^{\infty} P[\lambda_2(D(0,a_n)) \geq c\, k_2(a_n)] < \infty,$$

from which (3.*h*) follows immediately, with $C_2 = c$.

(3.3) Proof of (3.*d*). We cannot use *proposition 5* here. However we observe that the joint distribution of $(B(a+t)\ ;\ 0 \leq t \leq b-a)$ and $(B(c+s)\ ;\ 0 \leq s \leq d-c)$ is absolutely continuous with respect to that of two independent Brownian motions. Thus it suffices to prove the following statement. Let B, B' be two independent Brownian motions in $\mathbb{R}^3$, and let λ_2 be as above. If I denotes the set of intersection points of the paths of B and B', there exists a constant C_2' such that, a.s.,

$$(3.i) \qquad k_2 - m(\{y \in I\ ;\ \limsup_{a \to 0} \left(\frac{\lambda_2(D(y,a))}{k_2(a)}\right) \leq C_2'\}) = 0.$$

A scaling argument shows that we may replace I by $I \cap D(0,1)$ in (3.*i*). For each integer $n \geq 1$, we denote by Ω_n the set of cubes of $\mathbb{R}^3$ of the following type

$$A = \prod_{i=1}^{3} [k_i\, a_n\ ;\ (k_i+1)a_n],$$

where the k_i's are integers such that $-2^n \leq k_i < 2^n$ $(i = 1,2,3)$. Let N_n denote the number of cubes in Ω_n which are hit by both B and B'. It follows easily from the estimates in [9] that :

(3.j) $E[N_n] \leq c\ 2^n$,

for some constant c and any $n \geq 1$.

We now fix $n_o \geq 1$. For any $n > n_o$, let N'_n denote the number of cubes A belonging to Ω_n which are hit by both B and B', and such that, for any $k \in \{n_o, n_o+1, \ldots, n\}$,

$$\lambda_2(D(y_A, a_n)) \leq \gamma\ k_2(a_n).$$

Here y_A denotes the center of the cube A, and γ is a positive constant. We will first prove that, if γ is small enough, then a.s.,

(3.k) $\liminf_{n \to \infty} k_2(a_n) N'_n = 0.$

For any $A \in \Omega_n$, let $\mathfrak{J}_A$ denote the event {B and B' hit A}. Then,

(3.ℓ) $$E[N'_n] = \sum_{A \in \Omega_n} P[\mathfrak{J}_A] \times P\Big[\frac{\lambda_2(D(y_A, a_k))}{k_2(a_k)} \leq \gamma \text{ for any } k = n_o, \ldots, n / \mathfrak{J}_A\Big].$$

In order to bound the above conditional expectation, we will apply the Markov property at some suitable stopping times. We fix $A \in \Omega_n$, and set :

$$T_o = \inf\{t\ ;\ B_t \in A\},\quad T'_o = \inf\{t\ ;\ B'_t \in A\}.$$

Conditionally on $\mathfrak{J}_A$, we may define the following stopping times : for any $k = 1,2,\ldots,n-n_o$,

$$T_k = \inf\{t > T_o\ ;\ |B_t - y_A| = 2^{-(n-k)}\}$$

$$T'_k = \inf\{t > T'_o\ ;\ |B'_t - y_A| = 2^{-(n-k)}\}$$

(note that $A \subset D(y_A, a_n)$). As usual, let β_2 denote the intersection local time of B and B', we observe that, for any $k = 1,2,\ldots,n-n_o$,

$$(3.m) \quad \lambda_2(D(y_A, a_{n-k})) \geq \beta_2([T_{k-1} \,;\, T_k] \times [T'_{k-1} \,;\, T'_k]).$$

Let $(\mathcal{F}_t, t \geq 0)$, resp. $(\mathcal{F}'_t, t \geq 0)$, denote the canonical filtration of B, resp. B'. We now apply *proposition* 3 and *lemma* 4 , together with a suitable change of scale, to bound the probability of small values of the right member of (3.*m*). The Markov property will show that our bounds do not depend on the past of B, resp. B', up to T_{k-1}, resp. T'_{k-1}. The precise statement is as follows. Conditionally on $\mathfrak{J}_A$, for any $k = 1,2,\ldots,n-n_o$ and any $u > 0$ large enough,

$$(3.n) \quad P\left[\beta_2([T_{k-1} \,;\, T_k] \times [T'_{k-1} \,;\, T'_k]) \leq a_{n-k} \cdot u / \mathcal{F}_{T_{k-1}} \vee \mathcal{F}'_{T'_{k-1}}\right] \leq 1 - \exp(-\rho\, u^{1/2})$$

for some constant $\rho > 0$. Obviously ρ does not depend on A. It now follows from (3.*m*) and (3.*n*) that, if n_o is large enough,

$$(3.o) \quad P\left[\bigcap_{k=n_o}^{n} \left\{\frac{\lambda_2(D(y_A, a_k))}{k_2(a_k)} \leq \gamma\right\} / \mathfrak{J}_A\right] \leq \prod_{k=n_o}^{n-1} (1 - \exp(-\rho\, \gamma^{1/2}(\log k + \log\log 2))).$$

We take γ so small that $\rho\, \gamma^{1/2} < 1$. Then (3.*o*) implies

$$P\left[\bigcap_{k=n_o}^{n} \left\{\frac{\lambda_2(D(y_A, a_k))}{k_2(a_k)} \leq \gamma\right\} / \mathfrak{J}_A\right] \leq c' \exp(-c''(n-n_o)^{1-\rho\, \gamma^{1/2}}),$$

for some positive constants c', c''. Coming back to (3.*ℓ*) and using (3.*j*), we get

$$E[N'_n] \leq cc'\, 2^n \exp(-c''(n-n_o)^{1-\rho\, \gamma^{1/2}})$$

and (3.*k*) follows through an application of Fatou's lemma.

Now that we have established (3.*k*) it is a trivial matter to complete the proof of (3.*i*). Using (3.*k*) together with the definition of a Hausdorff measure, we obtain that, for γ small enough, a.s.,

$$k_2 - m(\{y \in I \cap D(0\ ,\ 1)\ ;\ \frac{\lambda_2(D(y,a_k))}{k_2(a_k)} \leq \gamma \quad \text{for any } k \geq n_o\}) = 0.$$

Since n_o was arbitrary, (3.*i*) follows.

4. PROOF OF THEOREM 1.

(4.1) Throughout this section, $B = (B_t\ ;\ t \geq 0)$ denotes a Brownian motion in $\mathbb{R}^2$ and $p \geq 2$ is an integer. We consider a subset J of $\mathcal{C}_p$ of the following type :

(4.*a*) $\quad J =]u_1\ ;\ v_1[\ \times\ ]u_2\ ;\ v_2[\ \times\dots\times\ ]u_p\ ;\ v_p[,$

where $0 < u_1 < v_1 < \dots < u_p < v_p$. We use the notations of subsection 2.2. The method described in subsection 3.1 allows us to reduce the proof of *theorem 1* to the existence of two positive constants C_1, C_1' (not depending on J) such that, a.s.,

(4.*b*) $$\ell_p^J(\{y \in \mathbb{R}^2\ ;\ \limsup_{a \to 0}(\frac{\ell_p^J(D(y,a))}{h_p(a)}) \geq C_1\}) = 0,$$

(4.*c*) $$h_p - m(\{y \in D_p^J\ ;\ \limsup_{a \to 0}(\frac{\ell_p^J(D(y,a))}{h_p(a)}) \leq C_1'\}) = 0.$$

The proof of (4.*c*) is similar to that of (3.*d*), but the proof of (4.*b*) is signicantly more difficult than that of (3.*c*). Therefore we will first prove (4.*c*).

(4.2) Proof of (4.*c*). The same arguments as in (3.3) show that it is enough to prove the following statement. Let $B^1,\dots,B^p$ be p independent Brownian motions in $\mathbb{R}^2$, and, for each $i = 1,\dots,p$ set $T^i = \inf\{t\ ;\ |B_t^i| = 1\}$. Let I denote the set of intersection points of the paths of B^i on $[0\ ;\ T^i]$ (for $i = 1,\dots,p$), and let λ_p^o be the measure on I which is the image of the intersection local time of $B^1,\dots,B^p$, restricted to the product $[0\ ;\ T^1] \times\dots\times [0\ ;\ T^p]$. Then,

$$(4.d)\qquad h_p - m(\{y \in I \cap D(0,1/2)\ ;\ \limsup_{a \to 0}(\frac{\lambda_p^o(D(y,a))}{h_p(a)}) \leq C_1'\}) = 0,$$

for some constant C_1'. In contrast with the case $d = 3$, note that we must limit ourselves to p-uples $(t_1,\dots,t_p)$ belonging to a product of finite intervals, because of the recurrence of planar Brownian motion.

In order to prove *(4.d)* we will use, instead of the sequence (a_n) of section 3, the sequence (b_n) defined by :

$$b_n = 2^{-2^n} \qquad (n \geq 1).$$

For any $n \geq 1$, let Δ_n denote the set of squares of $\mathbb{R}^2$ of the type :

$$A = [k_1\, b_n\ ;\ (k_1+1)b_n] \times [k_2\, b_n\ ;\ (k_2+1)b_n],$$

where k_1,k_2 are two integers such that : $-2^{2^n-1} \leq k_i < 2^{2^n-1}$ $(i = 1,2)$. Let N_n be the number of squares in Δ_n which are hit by each process B^i before time T^i. The estimates of [9] yield that :

$$E[N_n] \leq c\ 2^{2^{n+1}}\ 2^{-pn},$$

for some constant c. As in section 3 we denote by y_A the center of the square A. Fix $n_o \geq 1$ and for any $n > n_o$, let N_n' be the number of squares A in Δ_n which are hit by each process B^i before time T^i, and such that for any $k = n_o,\dots,n$,

$$\lambda_p^o(D(y_A,b_k)) \leq \gamma\ h_p(b_k)$$

where γ is some positive constant to be fixed later.

We now fix A and introduce the same notations as in section 3 : for any $i = 1,\dots,p$,

$$T_o^i = \inf\{t\ ;\ B_t^i \in A\}$$

and, for $k = 1,2,\dots,n-n_o$,

$$T_k^i = \inf\{t > T_o^i\ ;\ |B_t^i - y_A| = b_{n-k}\}.$$

Let $\mathcal{F}^i$ denote the canonical filtration of B^i, and let β_p be the intersection local time of $B^1,\dots,B^p$. Using *proposition 3* and *lemma 4*, and arguing as in subsection (3.3) it follows that for any $u > 0$ large enough, and for $k = 1,2,\dots,n-n_o$,

$$(4.e) \quad P\Big[\beta_p\Big(\prod_{i=1}^{p}\ ([T^i_{k-1}\ ;\ T^i_k]\cap\{s;|B^i_s-y_A|\leq b_{n-k+1}\})\Big)\leq u b^2_{n-k+1} \Big/ \bigvee_{i=1}^{p} \mathcal{F}^i_{T^i_{k-1}}\Big]$$

$$\leq 1 - \exp(-\rho\ |\log b_{n-k+1}|^{-1}\ u^{1/p}),$$

for some positive constant ρ. Now note that, conditionally on the event $\mathfrak{J}_A = \{T^i_o < T^i\ ;\ i = 1,\dots,p\}$,

$$\lambda^o_p(D(y_A,b_{n-k+1})) \geq \beta_p\Big(\prod_{i=1}^{p}\ ([T^i_{k-1}\ ;\ T^i_k]\cap\{s;|B^i_s-y_A| < b_{n-k+1}\})\Big)$$

for any $k = 1,2,\dots,n-n_o$. Hence it follows from (4.e) that, for n_o large enough,

$$P\Big[\bigcap_{k=n_o}^{n}\ \{\lambda^o_p(D(y_A,b_k)) \leq \gamma\ h_p(b_k)\}\ /\ \mathfrak{J}_A\Big]$$

$$\leq \prod_{k=n_o}^{n-1}\ (1 - \exp(-\rho\ \gamma^{1/p}\ \log\log|\log b_k|))$$

$$\leq c'\ \exp(-c''(n-n_o)^{1-\rho\ \gamma^{1/p}}),$$

if γ is small enough. As in section 3, we conclude that :

$$\lim_{n\to\infty} h_p(b_n)\ E[N'_n] = 0,$$

hence :

$$\liminf\ h_p(b_n)\ N'_n = 0,\ \text{a.s.},$$

which implies

$$h_p - m(\{y\in I\cap D(0,1/2)\ ;\ \lambda^o_p(D(y,b_k))\leq\gamma\ h_p(b_k)\ \text{for}\ k\geq n_o\}) = 0,$$

from which (4.d) follows easily.

(4.3) Proof of (4.c). Using *proposition* 5 the proof of (4.c) reduces to that of the following statement. Let $B^1,\dots,B^p$ be p independent Brownian motions in $\mathbb{R}^2$, starting from 0, and set

$$T^i = \inf\{t \ ; \ |B_t^i| = 1\}.$$

Let β_p denote the intersection local time of $B^1,\dots,B^p$ and let λ_p be the image measure on $\mathbb{R}^2$ of the restriction of β_p to $\Pi[0 \ ; \ T^i]$. Then there exists a constant C such that :

$$(4.f) \qquad \limsup_{a \to 0} \frac{\lambda_p(D(0,a))}{h_p(a)} \leq C \qquad \text{a.s..}$$

It is worth noting that the method we used in (3.2) only yields the weaker result :

$$\limsup_{k \to \infty} \frac{\lambda_p(D(0,b_k))}{h_p(b_k)} \leq C \qquad \text{a.s..}$$

Thus we will have to use another method, inspired by Ray's paper [15].

Let us fix a such that $0 < a < 1/4$. We introduce the following stopping times, for $i = 1,\dots,p$,

$$T_o^i = 0$$

$$S_o^i = \inf\{t \ ; \ |B_t^i| \geq 2a\}$$

and, by induction, for any $j \geq 1$,

$$T_j^i = \inf\{t > S_{j-1}^i \ ; \ |B_t^i| \leq a\}$$

$$S_j^i = \inf\{t > T_j^i \ ; \ |B_t^i| \geq 2a\}.$$

Set $N_i = N_i(a) = |\{j \geq 1 \ ; \ T_j^i \leq T^i\}|$.

Lemma 7. *There exist two positive constants* A *and* C *such that, for any* $a \in \,]0 \ ; \ 1/4[$,

$$E[(a^{-2}\lambda_p(D(0,a)) - A\, N_1 \dots N_p)^4] \leq C(\log \tfrac{1}{a})^{4p-2}.$$

We first assume the result of *lemma 7* and complete the proof of *(4.f)*. We apply *lemma 7* with $a = a_n = 2^{-n}$ and denote $N_i(a_n)$ by N_i^n. It follows that :

$$\sum_{n=1}^{\infty} E\Big[\big(a_n^{-2}(\log \frac{1}{a_n})^{-p}\ \lambda_p(D(0,a_n)) - A(\log \frac{1}{a_n})^{-p}\ N_1^n \dots N_p^n\big)^4\Big] < \infty,$$

which implies that :

$$\lim_{n\to\infty} (a_n^{-2}(\log \frac{1}{a_n})^{-p}\ \lambda_p(D(0,a_n)) - A(\log \frac{1}{a_n})^{-p}\ N_1^n \dots N_p^n) = 0, \text{ a.s..}$$

Hence,

$$(4.g) \qquad \limsup_{n\to\infty} \frac{\lambda_p(D(0,a_n))}{h_p(a_n)} = \limsup_{n\to\infty} \frac{A\ N_1^n \dots N_p^n}{(\log \frac{1}{a_n} \log\log\log \frac{1}{a_n})^p}\ .$$

According to Ray [15] (p. 441) we have, for $i = 1,\dots,p$,

$$(4.h) \qquad \limsup_{n\to\infty} \frac{N_i^n}{\log \frac{1}{a_n} \log\log\log \frac{1}{a_n}} \le \overline{C}, \qquad \text{a.s.,}$$

for some constant $\overline{C}$. *(4.f)* now follows from *(4.g)* and *(4.h)*.

Proof of *lemma 7*. We first introduce a few notations. For $i = 1,\dots,p$ and for any $j \ge 0$ we set :

$$\zeta_j^i = S_j^i - T_j^i \quad \text{and, for} \quad 0 \le t \le \zeta_j^i,\ \omega_j^i(t) = B^i(T_j^i + t).$$

Then,

$$(4.i) \qquad \lambda_p(D(0,a)) = \sum_{j_1=0}^{N_1} \dots \sum_{j_p=0}^{N_p} \beta_p\Big(\prod_{i=1}^{p} [T_{j_i}^i\ ;\ S_{j_i}^i]$$

$$\cap \{(s_1,\dots,s_p)\ ;\ B_{s_1}^1 \in D(0,a)\}\Big) = \sum_{j_1=0}^{N_p} \dots \sum_{j_p=0}^{N_p} \psi_a(\omega_{j_1}^1, \omega_{j_2}^2, \dots, \omega_{j_p}^p),$$

for a certain functional ψ_a : $\psi_a(\omega^1,\dots,\omega^p)$ is the total mass of the measure induced by the intersection local time of $\omega^1,\dots,\omega^p$ on the disk $D(0,a)$. Note that the common distribution of the N_i's is given by :

$$(4.j) \qquad P[N_i = k] = (\log 2/\log a)(1 - \log 2/\log a)^k \qquad (k \ge 0).$$

Starting from (4.i) easy estimates yield :

$$E\Big[(\lambda_p(D(0,a)) - \sum_{j_1=1}^{N_1} \cdots \sum_{j_p=1}^{N_p} \psi_a(\omega^1_{j_1},\ldots,\omega^p_{j_p}))^4\Big]^{1/4}$$

$$\leq p\; E\Big[(\sum_{j_2=0}^{N_2} \cdots \sum_{j_p=0}^{N_p} \psi_a(\omega^1_o,\omega^2_{j_2},\ldots,\omega^p_{j_p}))^4\Big]^{1/4}$$

$$\leq p\; E[(N_2+1)\ldots(N_p+1)]E[\psi_a(\omega^1_o,\omega^2_o,\ldots\omega^p_o)^4]^{1/4}$$

$$\leq C\, a^2(\log 1/a)^{p-1},$$

for some constant C. In the above bounds we implicitly used the fact that :

$$E[\psi_a(\omega^1_o,\omega^2_{j_2},\ldots,\omega^p_{j_p})^4] \leq E[\psi_a(\omega^1_o,\omega^2_o,\ldots,\omega^p_o)^4] = c\cdot a^2.$$

Therefore it suffices to prove that

$$(4.k)\qquad E\Big[(a^{-2} \sum_{j_1=1}^{N_1} \cdots \sum_{j_p=1}^{N_p} \psi_a(\omega^1_{j_1},\ldots,\omega^p_{j_p}) - A\, N_1\ldots N_p)^4\Big]$$
$$\leq C\, (\log 1/a)^{4p-2},$$

where we take $A = a^{-2}\, E[\psi_a(\omega^1_1,\ldots,\omega^p_1)]$. Scaling arguments show that A does not depend on a. Conditioning with respect to $N_1,\ldots,N_p$ and using (4.j), we observe that (4.k) follows from the bound : for any $n_1,\ldots,n_p \geq 1$:

$$(4.\ell)\qquad E\Big[(a^{-2} \sum_{j_1=1}^{n_1} \cdots \sum_{j_p=1}^{n_p} \psi_a(\omega^1_{j_1},\ldots,\omega^p_{j_p}) - An_1\ldots n_p)^4/N_1 = n_1,\ldots,N_p = n_p\Big]$$

$$\leq C(n_1^2\, n_2^4\ldots n_p^4 + n_1^4\, n_2^2 \ldots n_p^4 + \ldots + n_1^4 \ldots n_{p-1}^4\, n_p^2).$$

We will now prove (4.ℓ). From now on we will fix $n_1,\ldots,n_p \geq 1$, and the following assertions will hold <u>conditionally</u> on the event $\{N_1 = n_1,\ldots,N_p = n_p\}$. First of all, for any $i = 1,\ldots,p$, the conditional law of $(\omega^i_1,\omega^i_2,\ldots,\omega^i_{n_i})$ can be described as follows :

• ω_1^i starts with the uniform distribution $u_a(dx)$ on the circle $S(0,a)$ and the law of $(\omega_1^i(t), t \leq \zeta_1^i)$ is that of a Brownian motion killed when it exits $D(0,2a)$

• for any $j = 2,\dots,n_i$, $\omega_j^i(0)$ is conditionally independent of $\omega_1^i,\dots,\omega_{j-1}^i$ knowing $\omega_{j-1}^i(\zeta_{j-1}^i)$, and

$$P[\omega_j^i(0) \in dx/\omega_{j-1}^i(\zeta_{j-1}^i)] = \mu_a(\omega_{j-1}^i(\zeta_{j-1}^i),dx),$$

we shall not need the exact form of the kernel μ_a, but we observe here that there exists a constant $\gamma > 0$, not depending on a, such that , for any $y \in S(0,2a)$,

$$(4.m) \qquad \mu_a(y,dx) \geq \gamma\, u_a(dx)$$

• for any $j = 2,\dots,n_i$, ω_j^i is conditionally independent of $\omega_1^i,\dots\omega_{j-1}^i$ knowing $\omega_j^i(0)$, and its conditional distribution is that of a Brownian motion killed when it exits $D(0,2a)$.

We also observe that $(\omega_1^i,\dots,\omega_{n_i}^i)$ form a stationary sequence whose distribution is the same as that of $(\omega_{n_i}^i,\omega_{n_i-1}^i,\dots,\omega_1^i)$.

We denote by $\mathcal{F}^{k,p}$ the σ-field generated by the ω_j^i's $(k \leq i \leq p,\ 1 \leq j \leq n_i)$. We first bound :

$$(4.n) \qquad E\Big[(a^{-2} \sum_{j_1=1}^{n_1} \dots \sum_{j_p=1}^{n_p} \psi_a(\omega_{j_1}^1,\dots,\omega_{j_p}^p) - E\big[a^{-2} \sum_{j_1=1}^{n_1} \dots \sum_{j_p=1}^{n_p} \psi_a(\omega_{j_1}^1,\dots,\omega_{j_p}^p)/\mathcal{F}^{2,p}\big])^4\Big]$$

$$= E\Big[(\sum_{j_1=1}^{n_1} \Phi_a(\omega_{j_1}^1) - E\big[\sum_{j_1=1}^{n_1} \Phi_a(\omega_{j_1}^1)/\mathcal{F}^{2,p}\big])^4\Big]$$

where we have set for simplicity :

$$\Phi_a(\omega^1_{j_1}) = a^{-2} \sum_{j_2=1}^{n_2} \cdots \sum_{j_p=1}^{n_p} \psi_a(\omega^1_{j_1}, \omega^2_{j_2}, \ldots, \omega^p_{j_p}).$$

Then,

$$(4.o)\qquad E\Big[\Big(\sum_{j_1=1}^{n_1} \Phi_a(\omega^1_{j_1}) - E\Big[\sum_{j_1=1}^{n_1} \Phi_a(\omega^1_{j_1})/\mathcal{F}^{2,p}\Big]\Big)^4\Big]$$

$$= \sum_{j_1,j_1',j_1'',j_1'''} E\Big[(\Phi_a(\omega^1_{j_1}) - E[\Phi_a(\omega^1_{j_1})/\mathcal{F}^{2,p}])\ldots$$

$$\ldots(\Phi_a(\omega^1_{j_1'''}) - E[\Phi_a(\omega^1_{j_1'''})/\mathcal{F}^{2,p}])\Big]$$

Fix j_1,j_1',j_1'',j_1''' such that $j_1 < j_1' \leq j_1'' < j_1'''$ and consider

$$E\Big[(\Phi_a(\omega^1_{j_1'''}) - E[\Phi_a(\omega^1_{j_1'''})/\mathcal{F}^{2,p}])\ /\mathcal{F}^{2,p} \vee \mathcal{F}^1_{j_1''}\Big],$$

where $\mathcal{F}^1_{j_1''}$ denotes the σ-field generated by the ω^1_j's $(1 \leq j \leq j_1'')$. Using our previous remarks we way replace the latter expression by :

$$\int \mu_a^{(j_1'''-j_1'')}(\omega^1_{j_1''}(\zeta^1_{j_1''}),dw)$$

$$\times\ E\Big[(\Phi_a(\omega^1_{j_1'''}) - E[\Phi_a(\omega^1_{j_1'''})/\mathcal{F}^{2,p}])/\mathcal{F}^{2,p} \vee \{\omega^1_{j_1'''}(0) = w\}\Big]$$

where $\mu_a^{(j_1'''-j_1'')}(y,dw)$ denotes the conditional distribution of $\omega^1_{j_1'''}(0)$ knowing that $\omega^1_{j_1''}(\zeta^1_{j_1''}) = y$ (with our previous notations $\mu_a^{(1)}(y,dw) = \mu_a(y,dw)$). It should be clear from *(4.m)* and the above remarks that for any y,

$$\mu_a^{(j_1'''-j_1'')}(y,dw) \geq (1-(1-\gamma)^{j_1'''-j_1''})u_a(dw).$$

Set $\nu_a^{(j_1'''-j_1'')}(y,dw) = \mu_a^{(j_1'''-j_1'')}(y,dw) - (1-(1-\gamma)^{j_1'''-j_1''})u_a(dw)$.
Noting that

$$\int u_a(dw)E\Big[(\Phi_a(\omega^1_{j_1'''}) - E[\Phi_a(\omega^1_{j_1'''})/\mathcal{F}^{2,p}])/\mathcal{F}^{2,p} \vee \{\omega^1_{j_1'''}(0) = w\}\Big] = 0,$$

we find,

$$(4.p)\quad E\Big[(\Phi_a(\omega^1_{j_1'''}) - E[\Phi_a(\omega^1_{j_1'''})/\mathcal{F}^{2,p}])\ /\mathcal{F}^{2,p} \vee \mathcal{F}^1_{j_1''}\Big]$$

$$= \int \nu_a^{(j_1'''-j_1'')}(\omega^1_{j_1''}(\zeta^1_{j_1''}),dw)$$

$$\times E\Big[(\Phi_a(\omega^1_{j_1'''}) - E[\Phi_a(\omega^1_{j_1'''})/\mathcal{F}^{2,p}])\ /\ \mathcal{F}^{2,p} \vee \{\omega^1_{j_1'''}(0) = w\}\Big].$$

Let $\bar{\mathcal{F}}^1_{j_1'}$ denote the σ-field generated by the ω^1_j's , $j_1' \leq j \leq n_1$.

Using the same arguments as above we obtain :

$$(4.q)\quad E\Big[(\Phi_a(\omega^1_{j_1}) - E[\Phi_a(\omega^1_{j_1})/\mathcal{F}^{2,p}])/\mathcal{F}^{2,p} \vee \bar{\mathcal{F}}^1_{j_1'}\Big]$$

$$= \int \bar{\nu}_a^{(j_1'-j_1)}(\omega^1_{j_1'}(0),dw)$$

$$\times E\Big[(\Phi_a(\omega^1_{j_1}) - E[\Phi_a(\omega^1_{j_1})/\mathcal{F}^{2,p}])/\mathcal{F}^{2,p} \vee \{\omega^1_{j_1}(\zeta^1_{j_1}) = w\}\Big]$$

where, for any $y \in S(0,a)$, $\bar{\nu}_a^{(j_1'-j_1)}(y,dw)$ is some positive measure on $S(0,2a)$, whose total mass is bounded by $(1-\gamma)^{j_1'-j_1}$. It follows that, whenever $j_1 < j_1' \leq j_1'' < j_1'''$,

$$\Big|E\Big[(\Phi_a(\omega^1_{j_1}) - E[\Phi_a(\omega^1_{j_1})/\mathcal{F}^{2,p}])\ldots(\Phi_a(\omega^1_{j_1'''}) - E[\Phi_a(\omega^1_{j_1'''})/\mathcal{F}^{2,p}])\Big]\Big|$$

$$= \Big|E\Big[E\Big[(\Phi_a(\omega^1_{j_1}) - E[\Phi_a(\omega^1_{j_1})/\mathcal{F}^{2,p}])/\mathcal{F}^{2,p} \vee \bar{\mathcal{F}}^1_{j_1'}\Big]$$

$$\times\ (\Phi_a(\omega^1_{j_1'}) - E[\Phi_a(\omega^1_{j_1'})/\mathcal{F}^{2,p}])(\Phi_a(\omega^1_{j_1''}) - E[\Phi_a(\omega^1_{j_1''})/\mathcal{F}^{2,p}])$$

$$\times\ E\Big[(\Phi_a(\omega^1_{j_1'''}) - E[\Phi_a(\omega^1_{j_1'''})/\mathcal{F}^{2,p}])/\mathcal{F}^{2,p} \vee \mathcal{F}^1_{j_1''}\Big]\Big]\Big|$$

$$\leq (1-\gamma)^{(j_1'-j_1)/2}\ (1-\gamma)^{(j_1'''-j_1'')/2}$$

$$\times\ E\Big[(\Phi_a(\omega^1_{j_1}) - E[\Phi_a(\omega^1_{j_1})/\mathcal{F}^{2,p}])^2\ldots$$

$$\ldots(\Phi_a(\omega^1_{j_1'''}) - E[\Phi_a(\omega^1_{j_1'''})/\mathcal{F}^{2,p}])^2\Big]^{1/2},$$

by Cauchy - Schwarz inequality. Here we use identities (4.p) and (4.q) together with the fact that the total mass of $\nu_a^{(j_1'''-j_1'')}(y,\cdot)$, resp. $\bar{\nu}_a^{(j_1'-j_1)}(y,\cdot)$, is bounded by $(1-\gamma)^{j_1'''-j_1''}$, resp. $(1-\gamma)^{j_1'-j_1}$. Trivial estimates yield :

$$E\Big[(\Phi_a(\omega^1_{j_1}) - E[\Phi_a(\omega^1_{j_1})/\mathcal{F}^{2,p}])^2 \dots (\Phi_a(\omega^1_{j_1'''}) - E[\Phi_a(\omega^1_{j_1'''})/\mathcal{F}^{2,p}])^2\Big]^{1/2}$$

$$\leq C\, n_2^4 \dots n_p^4,$$

hence , summing over all 4-tuples j_1, j_1', j_1'', j_1''' ,

$$\Big| \sum_{\substack{j_1,j_1',j_1'',j_1''' \\ j_1<j_1'\leq j_1''<j_1'''}} E\Big[(\Phi_a(\omega^1_{j_1}) - E[\Phi_a(\omega^1_{j_1})/\mathcal{F}^{2,p}]) \dots$$

$$\dots (\Phi_a(\omega^1_{j_1'''}) - E[\Phi_a(\omega^1_{j_1'''})/\mathcal{F}^{2,p}])\Big]\Big|$$

$$\leq C'\, n_1^2\, n_2^4 \dots n_p^4,$$

where $C' = C\cdot(1-(1-\gamma)^{1/2})^{-2}$. Similar arguments allow us to deal with other orderings of j_1, j_1', j_1'', j_1'''. Taking *(4.o)* into account, we get :

$$E\Big[(\sum_{j_1=1}^{n_1} \Phi_a(\omega^1_{j_1}) - E[\sum_{j_1=1}^{n_1} \Phi_a(\omega^1_{j_1})/\mathcal{F}^{2,p}])\Big]^4$$

$$\leq C''\, n_1^2\, n_2^4 \dots n_p^4,$$

for some constant C''. For simplicity set

$$U_a = \sum_{j_1=1}^{n_1} \Phi_a(\omega^1_{j_1}) = a^{-2} \sum_{j_1=1}^{n_1} \dots \sum_{j_p=1}^{n_p} \psi_a(\omega^1_{j_1},\dots,\omega^p_{j_p}),$$

so that we have proved

$$E[(U_a - E[U_a/\mathcal{F}^{2,p}])^4] \leq C''\, n_1^2\, n_2^4 \dots n_p^4.$$

The same arguments as above suffice to prove, for any $i = 1,\dots,p$

$$(4.r) \qquad E\Big[([U_a/\mathcal{F}^{i,p}] - E[U_a/\mathcal{F}^{i+1,p}])^4\Big] \leq C''\, n_1^4 \dots n_{i-1}^4\, n_i^2\, n_{i+1}^4 \dots n_p^4$$

where we agree that $\mathcal{F}^{p+1,p}$ is the trivial σ-field. The bound *(4.l)* follows from *(4.r)* and Minkowski inequality. □

Remark : The above arguments are adapted from Ray's proof in the special case p = 1. In this case it is possible to give a very simple proof of (4.b), using stochastic calculus : see [10]. Unfortunately the arguments of [10] do not apply to the general case.

REFERENCES

[1] Z. CIESIELSKI and S.J. TAYLOR. First passage times and sojourn times for Brownian motion in space and the exact Hausdorff measure of the sample path. Trans. Amer. Math. Soc. 103, 434-450 (1962).

[2] A. DVORETZKY, P. ERDÖS and S. KAKUTANI. Double points of paths of Brownian motion in n-space. Acta Sci. Math. (Szeged) 12, 64-81 (1950).

[3] A. DVORETZKY, P. ERDÖS and S. KAKUTANI. Multiple points of paths of Brownian motion in the plane. Bull. Res. Council Isr. Sect. F 3, 364-371 (1954).

[4] A. DVORETZKY, P. ERDÖS, S. KAKUTANI and S.J. TAYLOR. Triple points of Brownian motion in 3-space. Proc. Camb. Philos. Soc. 53, 856-862 (1957).

[5] E.B. DYNKIN. Additive functionals of several time-reversible Markov processes. J. Funct. Anal. 42, 64-101 (1981).

[6] E.B. DYNKIN. Random fields associated with multiple points of the Brownian motion. J. Funct. Anal. 62, 397-434 (1985).

[7] B. FRISTEDT. An extension of a theorem of S.J. Taylor concerning the multiple points of the symmetric stable process. Z. Wahrsch. verw. Gebiete 9, 62-64 (1967).

[8] D. GEMAN, J. HOROWITZ and J. ROSEN. A local time analysis of intersections of Brownian paths in the plane. Ann. Probab. 12, 86-107, (1984).

[9] J.F. LE GALL. Sur la saucisse de Wiener et les points multiples du mouvement brownien. Preprint (1984), to appear in Ann. Probab.

[10] J.F. LE GALL. Sur la mesure de Hausdorff de la courbe brownienne. In : Séminaire de Probabilités XIX. Lect. Notes Math. 1123, p. 297-313. Berlin, Heidelberg, New-York: Springer 1985.

[11] J.F. LE GALL. Sur le temps local d'intersection du mouvement brownien plan et la méthode de renormalisation de Varadhan. In : Séminaire de Probabilités XIX. Lect. Notes Math. 1123, p. 314-331. Berlin, Heidelberg, New-York, Springer 1985.

[12] J.F. LE GALL. Propriétés d'intersection des marches aléatoires, I. Convergence vers le temps local d'intersection. Comm. Math. Phys. 104, 471-507 (1986).

[13] J.F. LE GALL. Le comportement du mouvement brownien entre les deux instants où il passe par un point double. Preprint (1985), to appear in J. Funct. Anal.

[14] P. LEVY. La mesure de Hausdorff de la courbe du mouvement brownien. Giorn. Ist. Ital. Attuari, 16, 1-37 (1953).

[15] D. RAY. Sojourn times and the exact Hausdorff measure of the sample path for planar Brownian motion. Trans. Amer. Math. Soc. 106, 436-444 (1963).

[16] C.A. ROGERS, S.J. TAYLOR. Functions continuous and singular with respect to a Hausdorff measure. Mathematika 8, 1-31 (1961).

[17] J. ROSEN. Self-intersections of random fields. Ann. Probab. 12, 108-119 (1984).

[18] J. ROSEN. A local time approach to the self-intersections of Brownian paths in space. Comm. Math. Phys. 88, 327-338 (1983).

[19] S.J. TAYLOR. The exact Hausdorff measure of the sample path for planar Brownian motion. Proc. Camb. Philos. Soc. 60, 253-258 (1964).

[20] S.J. TAYLOR. Multiple points for the sample paths of the symmetric stable process. Z. Wahrsch. Verw. Gebiete 5, 247-264 (1966).

[21] S.J. TAYLOR. Sample path properties of processes with stationary independent increments. In : Stochastic Analysis, Kendall, D. and Harding, E. (eds). London, J. Wiley and Sons 1973.

Jean-François Le Gall

UNIVERSITE P. et M. CURIE
Laboratoire de Probabilités

4, place Jussieu - Tour 56

F - 75252 PARIS CEDEX 05

THE PACKING MEASURE OF PLANAR BROWNIAN MOTION

by

J-F. Le Gall and S. James Taylor

In a recent paper [4] a new fractal measure φ-p(E) with respect to the monotone function $\varphi(s)$ was defined, and it was shown that $\varphi_1(s) = s^2\left[\log\log\frac{1}{s}\right]^{-1}$ is the right function for measuring $B^d(t)$, the Brownian motion in $\mathbb{R}^d$, $d \geq 3$, in the sense that there are finite positive constants c_d such that

$$\varphi_1\text{-p } B^d(A) = c_d|A| \text{ a.s.} \tag{1}$$

for every Borel $A \subset \mathbb{R}^+ = [0,\infty)$, where $|A|$ denotes the Lebesgue measure of A. This new measure φ-p(E), which we called φ-packing measure, involves maximising the φ content of a packing by disjoint balls centered in E, with radii at most δ, and taking the limit as $\delta \downarrow 0$. In the present note we consider the problem of determining a function φ_2 which gives the analogue of (1) for the critical case of planar Brownian motion $B^2(t)$, which we will henceforth denote by Z(t). In fact, we show that no such φ_2 exists, and give a test which determines whether φ-p Z[0,1] is 0 or $+\infty$. This provides a complete solution to Problem 1 of [5].

The reader is referred to [4] for the precise definition of the measure φ-p and for the proofs of its main properties. In particular, we will need the following density theorem.

LEMMA 1. *If μ is a finite Borel measure in $\mathbb{R}^d$ and φ is a measure function, there is a positive λ such that for all Borel sets* $E \subset \mathbb{R}^d$

$$\lambda\mu(E) \inf_{x\in E} \left\{\limsup_{r\downarrow 0} \frac{\varphi(2r)}{\mu(B_r(x))}\right\} \leq \varphi\text{-p}(E)$$

$$\leq \mu(E) \sup_{x\in E} \left\{\limsup_{r\downarrow 0} \frac{\varphi(2r)}{\mu(B_r(x))}\right\}.$$

In the above, $B_r(x)$ denotes the closed ball of radius r centered at x. Let Γ denote the collection of semi-dyadic squares S in $\mathbb{R}^2$, that is squares of side 2^{-n} parallel to the axes for which each projection on an axis is of the form $[\frac{1}{2} k2^{-n}, (\frac{1}{2} k+1)2^{-n})$, $k \in \mathbb{Z}$; and let $\Gamma_m(E)$ be the subcollection of such squares S with $n \geq m$ and such that there is a point $x \in E$ whose distance from the complement of S is at least 2^{-n-2}. If we use disjoint squares of $\Gamma_m(E)$ to pack E, the resulting φ-packing measure is denoted $\varphi\text{-p}^{**}(E)$ and it differs from φ-p(E) by no more than a bounded factor. We use this in the following form. Given φ there is a finite constant K such that

$$h\text{-p}(E) \leq K \limsup_{n\to\infty} \{\textstyle\sum h(d(S_i)) : S_i \text{ disjoint } \in \Gamma_n(E)\} \tag{2}$$

As is usual in these arguments we will apply lemma 1 to the occupation measure μ_ω determined by the trajectory. Thus

$$\mu_\omega(A) = |\{t \in (0,1): Z(t) \in A\}| \tag{3}$$

defines a finite Borel measure in $\mathbb{R}^2$, which is spread evenly over the path $Z(0,1)$. Since $Z(t)$ spends time in the disc $B_r(x)$, with $x = Z(t_0)$ both before and after $t = t_0$, and we are interested in the small values of $\mu_\omega(B_r(x))$ as $r \downarrow 0$, we will need to add the random variables corresponding to $t \in (0,t_0)$ and $t \in (t_0,1)$. We will make a transformation which requires information about the Bes(4) process as $t \to \infty$, i.e., the process $\rho(t) = d(0,B^4(t))$, the radial component of Brownian motion in $\mathbb{R}^4$. This is given by

LEMMA 2. _If_ $X(t)$ _is the square of a_ Bes(4) _process_ $\rho(t)$, _then_

$$\lim_{u\to\infty} \frac{2e^{2u}}{X(u)} \int_u^\infty e^{-2y} X(y)\,dy = 1 \text{ a.s.}$$

PROOF: Fix $\alpha \in (\frac{1}{2},1)$ and take $a_k = k^2$ $(k \geq 1)$,

$$D_k = \{\omega: \sup[|X(t)-X(a_k)|;\ a_k < t \leq a_{k+1}] > k^\alpha\}.$$

Easy estimates give $\sum P(D_k) < \infty$ so that, a.s. there is a finite $k_0 = k_0(\omega)$ such that $\omega \notin D_k$ for $k \geq k_0$. On the other hand, using the classical law of iterated logarithm for $\rho(t)$, and the rate of escape for $B^4(t)$ established by Dvoretzky, Erdös [1] we see that a.s. there is a $t_0 = t_0(\omega)$ such that

$$t(\log t)^{-2} \le X(t) \le 3t \log\log t \qquad (4)$$

for $t \ge t_0$. Now suppose $u \ge t_0$ and $a_p < u \le a_{p+1}$, $p \ge k_0$. Then

$$|2e^{2u}\int_u^\infty e^{-2y}X(y)dy - X(u)|$$

$$\le X(u)(1-2e^{2u}\int_u^{a_{p+2}} e^{-2y}dy)$$

$$+ 2e^{2u}\int_u^{a_{p+2}} e^{-2y}|X(y)-X(u)|dy + 2e^{2u}\int_{a_{p+2}}^\infty e^{-2y}X(y)dy.$$

Dividing by $X(u)$, using (4) and the bound $|X(y)-X(u)| \le 2(p+1)^\alpha$ in the second term gives

$$\lim_{u\to\infty} \left[\frac{2e^{2u}}{X(u)} \int_u^\infty e^{-2y}X(y)dy - 1\right] = 0.$$

THEOREM 3. *Let* $Z_1(t)$, $Z_2(t)$ *be two independent Brownian motions in* $\mathbb{R}^2$, *with* $Z_1(0) = Z_2(0) = 0$ *and let*

$$T_i(r) = \int_0^1 1_{B_r(0)}(Z_i(s))ds \qquad i = 1,2.$$

If $g: \mathbb{R}^+ \longrightarrow \mathbb{R}^+$ *is monotone increasing, then*

$$\liminf_{r\downarrow 0} \frac{T_1(r)+T_2(r)}{r^2\log\frac{1}{r}\, g(r)} = \begin{matrix}0\\ \text{or}\\ \infty\end{matrix} \ \textit{according as}\ \sum g(2^{-2^k}) \begin{matrix}=\infty\\ \text{or.}\\ <\infty\end{matrix}$$

PROOF: Let $\sigma_i = \inf\{t > 0: |Z_i(t)| = 1\}$ $i = 1,2$. It suffices to prove the theorem with $T_i(r)$ replaced by

$$\tilde{T}_i(r) = \int_0^{\sigma_i} 1_{B_r(0)}(Z_i(s))ds.$$

Now put $Y_i(t) = |Z_i(t)|$, $i = 1,2$ to give two Bes(2) processes. Denote their local times at the times σ_i by $(\ell_x^i,\ x \geq 0)$. Then

$$\tilde{T}_i(r) = \int_0^r \ell_x^i\, dx \quad i = 1,2.$$

Using the first proposition of [2], there are processes $U_i(t)$ which have the same distribution as the square of a Bes(2) process such that

$$\ell_x^i = x\, U_i(\log \tfrac{1}{x}) \text{ for } x \in (0,1);\ i = 1,2.$$

Since the Y_i processes are independent it follows that U_1 and U_2 are independent which means that U_1+U_2 is the square of a Bes(4) process, denoted by X in Lemma 2. Thus

$$\tilde{T}_1(r) + \tilde{T}_2(r) = \int_0^r x(U_1+U_2)(\log \tfrac{1}{x})dx$$

$$= \int_{\log \frac{1}{r}}^{\infty} e^{-2y}X(y)dy$$

and we have converted the behaviour of $\tilde{T}_1 + \tilde{T}_2$ as $r \downarrow 0$ to that of process $X(y)$ as $y \to \infty$. By Lemma 2, we have

$$\lim_{r\downarrow 0} \frac{\tilde{T}_1(r)+\tilde{T}_2(r)}{2r^2X(\log \frac{1}{r})} = 1 \text{ a.s.}$$

But now the rate of escape for $B^4(t)$ as $t \to \infty$ tells us that

$$\liminf_{t\to\infty} \frac{X(t)}{t\, h(t)} = \begin{matrix} 0 \\ \text{or} \\ \infty \end{matrix} \text{ according as } \sum_{n=1}^{\infty} h(2^n) = \begin{matrix} \infty \\ \text{or.} \\ <\infty \end{matrix}$$

Putting $h(t) = g(2^{-t})$ now completes the proof.

REMARK: A simplified version of the above argument gives an alternative proof of Ray's result [3]

$$\limsup_{r\downarrow 0} \frac{T(r)}{r^2 \log \frac{1}{r} \log\log\log \frac{1}{r}} = 1 \quad \text{a.s.}$$

We can now state the main result of the present paper.

THEOREM 4. <u>Suppose</u> $Z(t)$ <u>is a planar Brownian motion and</u> $Z(0,1)$ <u>denotes the sample path up to time</u> 1. <u>If</u> $\varphi(s) = s^2 \log \frac{1}{s}\, h(s)$ <u>where</u> $h[0,1) \longrightarrow [0,1)$ <u>is monotone increasing but</u> $h(s) \log \frac{1}{s}$ <u>is decreasing, then</u> a.s.

$$\varphi\text{-}p\ Z(0,1) = \begin{matrix} 0 \\ \text{or} \\ \infty \end{matrix} \ \underline{\text{according as}} \ \sum h(2^{-2^k}) \begin{matrix} <\infty \\ \text{or} \\ =\infty \end{matrix}.$$

NOTE: The convergence of the series is equivalent to the convergence of the integral $\int_{0+} \frac{h(s)}{s \log \frac{1}{s}}\, ds$.

PROOF: First suppose that $\sum h(2^{-2^k})$ diverges. Then theorem 3 implies that

$$\limsup_{r\downarrow 0} \frac{r^2 \log \frac{1}{r}\, h(r)}{T_1(r)+T_2(r)} = +\infty \quad \text{a.s.}$$

Since $Z(t)-Z(t_0)$, $t \ge t_0$ and $Z(t_0-s)-Z(t_0)$, $s \ge 0$ behave like two independent Brownian motions starting at 0 it follows from (3) that, for each fixed $t_0 \in (0,1)$, a.s.

$$\limsup_{r \downarrow 0} \frac{\varphi(2r)}{\mu_\omega B_r(Z(t_0))} = +\infty, \tag{5}$$

and a Fubini argument shows that the set A of $t_0 \in (0,1)$ satisfying (5) has full Lebesgue measure a.s. which means that $\mu_\omega\, Z(A) = 1$. Lemma 1 applied to the occupation measure μ_ω now gives φ-p $Z(0,1) = +\infty$ a.s.

Now suppose $\sum h(2^{-2^k})$ converges. We cannot use lemma 1 because of the exceptional set of t_0 where

$$\limsup \frac{\varphi(2r)}{\mu_\omega\, B_r(Z(t_0))} = 0$$

may fail, but we can argue directly on the sample path. Apparently crude estimates are good enough because a planar path which hits a disc of radius r has large probability of getting within r^2 of every point of the disc within unit time (as $r \downarrow 0$), and the relevant measure functions φ are very close to $\varphi(s) = s^2$ so we do not lose more than a constant factor when we replace a packing square of side d by a large number of squares inside of side at least d^2.

It is clearly sufficient to replace $Z(0,1)$ by $Z(0,\sigma)$ where $\sigma = \inf\{t > 0: |Z(t)| = 1\}$. It is also sufficient to show that, for each fixed $\varepsilon > 0$, φ-p$(A_\varepsilon) = 0$ a.s. where

$$A_\varepsilon = Z(0,\sigma) \cap \{z \in \mathbb{R}^2: \varepsilon \le |z| \le 1\}.$$

Now let $\{S_i\}$ be a disjoint packing of A_ε by semi-dyadic squares of $\Gamma_n(A_\varepsilon)$. A typical S_i has side 2^{-n_i}, $n_i \ge n$, and we can choose the integer k_i such that $2^{k_i-1} < n_i \leq 2^{k_i}$. The conditions satisfied by h(s) imply that

$$\varphi(2^{-n_i}) \leq 2^{2(2^{k_i}-n_i)}\varphi(2^{-2^{k_i}})$$

and we note that $2^{2(2^{k_i}-n_i)}$ is the number of dyadic squares of side $2^{-2^{k_i}}$ contained in S_i. This means that we increase $\sum \varphi(d(S_i))$ by replacing each S_i by all the dyadic squares inside it of side $2^{-2^{k_i}}$. Since each $S_i \in \Gamma_n(A_\varepsilon)$, each of these smaller squares of side $2^{-2^{k_i}}$ is within $2\cdot 2^{-2^{k_i-1}}$ of A_ε. But the S_i are disjoint, so $\sum \varphi(d(S_i)) \leq \sum_{k=p}^{\infty} \{\sum \varphi(d(S))$: S has side 2^{-2^k} and $d(A_\varepsilon, S) \leq 2\cdot 2^{-2^{k-1}}\}$.
But for a fixed S of side 2^{-2^k} we can estimate the probability that S is within $2\cdot 2^{-2^{k-1}}$ of $Z(0,\sigma)$ by noting that this implies that Z(t) hits the disc of this radius with centre in S. This implies that

$$P(d(A_\varepsilon, S) < 2\cdot 2^{-2^{k-1}}) < K_\varepsilon 2^{-k}$$

for a constant depending on ε but not on S. This allows us to estimate the first moment of the φ content of all dyadic squares of side 2^{-k} which are close to Z(0,1).

$$E\left[\sum_{k=p}^{\infty} \Sigma\{\varphi(d(S)) : S \text{ of side } 2^{-2^k} \text{ and } d(A_\varepsilon,S) \le 2\cdot 2^{-2^{k-1}}\}\right]$$

$$\le 16 \sum_{k=p}^{\infty} \varphi(2^{-2^k})2^{2^{k+1}} K_\varepsilon 2^{-k}$$

$$\le c \sum_{k=p}^{\infty} h(2^{-2^k}) \longrightarrow 0 \text{ as } p \longrightarrow \infty$$

since this is the tail of a convergent series. It now follows from (2) that φ-p $Z(0,1) = 0$ a.s.

REFERENCES

1. A. Dvoretzky and P. Erdös. Some problems on random walk in space. Proc. Second Berkeley Symp. Berkeley (1950), 353-267.

2. J-F. Le Gall. Sur la mesure de Hausdorff de la courbe Brownienne. Séminaire de Probabilités XIX. Springer Lecture Notes 1123 (1985), 297-313.

3. D. Ray. Sojourn times and the exact Hausdorff measure of the sample path for planar Brownian motion. Trans. Amer. Math. Soc. 106 (1963), 436-444.

4. S. J. Taylor and C. Tricot. Packing measure and its evaluation for a Brownian path. Trans. Amer. Math. Soc. 288 (1985), 679-699.

5. S. J. Taylor. The use of packing measure in the analysis of random sets. Proc. of 15th Symposium on Stochastic Processes (1985).

Laboratoire de Probabilités
Université de Paris VI
4, Place Jussieu - Tour 56
75230 Paris FRANCE

Department of Mathematics
University of Virginia
Charlottesville
Virginia 22903 U.S.A.

TRUNCATED GAUGE AND SCHRÖDINGER OPERATOR WITH BOTH SIGH EIGENVALUES

by

ZHIMING MA and ZHONGXIN ZHAO

Let D be a bounded regular domain in R^d ($d \geqslant 1$), $q \in K_d^{loc}$ (see [2] for the definition) and $f \in C(\partial D)$. Consider the Dirichlet problem:

$$\begin{cases} \frac{\Delta}{2} u + qu = 0 \\ u/\partial D = f \end{cases} \quad \text{in } D \tag{1}$$

Let $\{X_t\}$ be d-dimensional Brownian motion. Define the semigroup of operators:

$$T_t \varphi(x) = E^x \left[t < \tau_D, \ e_q(t) \varphi(X_t) \right], \quad t > 0, \ x \in D,$$

where

$$e_q(t) = \exp \left[\int_0^t q(x_s) ds \right], \quad t \geqslant 0 .$$

Let $\{\lambda_k\}$ and $\{\varphi_k\}$ denote the eigenvalues and eigenfunctions of $\frac{\Delta}{2} + q$ with the Dirichlet boundary condition in D, respectively.

In this paper, we assume

$$\lambda_k \neq 0 \ , \ \forall k \geqslant 1 \ , \tag{2}$$

and give some probabilistic representations of the solution to problem (1).

Take a constant $\mu > \lambda_1$. Then by Theorem 4.9 in [1] we have

$$u^{\mu}(x) = E^{x}\left[e_{q-\mu}(\tau_D)\ f(X(\tau_D))\right] < \infty \text{ in } D\ ,$$

which gives the solution to the problem:

$$\begin{cases} \frac{\Delta}{z} u + qu - \mu u = 0 & \text{in } D \\ u/_{\partial D} = f\ , \end{cases} \tag{3}$$

where τ_D is the first exit time from D.

PROPOSITION 1. $\forall t > 0$,

$$u = u^{\mu} + \mu \int_0^t T_s ds\ (I - T_t)^{-1} u^{\mu} \tag{4}$$

is the solution to problem (1).

PROOF. By assumption (2), $\forall\ t > 0$, 1 is not an eigenvalue of T_t, hence $(I - T_t)^{-1}$ is a bounded operator in $L^2(D)$. Consequently, we have $(I - T_t)^{-1}\ u^{\mu} \epsilon L^2(D)$. By Theorem 3.12 in [1],

$$\mu \int_0^t T_s ds (I - T_t)^{-1} u^{\mu} \epsilon\ C_0(D) \cap \text{Domain [generator of } \{T_t\} \text{ in } L^2(D)],$$

where $C_0(D)$ is the space of continuous functions in D vanishing on the boundary. Thus,

$$u/_{\partial D} = u^{\mu}/_{\partial D} = f$$

and

$$\left(\frac{\Delta}{z} + q\right) u = \left(\frac{\Delta}{z} + q\right)u^{\mu} + \mu(T_t - 1)\ (I - T_t)^{-1}\ u^{\mu}$$

$$= \mu\ u^{\mu} - \mu u^{\mu} = 0\ .$$

□

PROPOSITION 2. $\forall\ t > 0$, u is the solution to problem (1) is and only if $\forall\ x\ \epsilon D$,

$$E^{x}\left[\ \tau \leqslant t,\ e_q(\tau)\, f(x_{\tau})\ \right] = (I - T_t)\, u(x)\ , \tag{5}$$

where $\tau = \tau_D$.

REMARK. Since T_t is an integral operator, (5) is an integral equation.

PROOF. We have $\forall t > 0$,

$$E^{\cdot}\left[\tau \leq t,\ e_{q-\mu}(\tau)\, f(x_\tau)\right] = u^{\mu} - E^{\cdot}\left[t < \tau,\ e_{q-\mu}(\tau)\ f(X_\tau)\right]$$

$$= u^{\mu} - E^{\cdot}\left\{ t < \tau,\ e_{q-\mu}(t)\, E^{x(t)}[e_{q-\mu}(\tau)\, f(x_\tau)]\right\}$$

$$= u^{\mu} - e^{-\mu t}\, T_t u^{\mu}\ . \tag{6}$$

Thus,

$$E^{\cdot}\left[\tau \leq t\cdot\ e_q(\tau)\, f(x_\tau)\right] = E^{\cdot}\left[\tau \leq t,\ e^{\mu t} e_{q-\mu}(\tau)\, f(x_\tau)\right]$$

$$= E^{\cdot}\left[\tau \leq t,\ (1 + \mu \int_0^{\tau} e^{\mu s} ds)\ e_{q-\mu}(\tau)\ f(x_\tau)\right]$$

$$= E^{\cdot}\left[\tau \leq t,\ e_{q-\mu}(\tau)\, f(x_\tau)\right]$$

$$+ \mu \int_0^t E^{\cdot}\left\{s < \tau,\ e^{\mu s} e_{q-\mu}(s)\, E^{x(s)}\left[\tau \leq t\text{-}s,\ e_{q-\mu}(\tau)\, f(x_\tau)\right]\right\} ds$$

$$= u^{\mu} - e^{-\mu t}\, T_t u^{\mu} + \mu \int_0^t T_s\ \left[u^{\mu} - e^{-\mu(t-s)}\ T_{t-s}\ u^{\mu}\right] ds$$

$$= u^{\mu} - T_t\, u^{\mu} + \mu \int_0^t T_s\, ds\, u^{\mu}\ , \tag{7}$$

where in the fourth equality we use (6) to the both terms.

Thus, Proposition 2 follows from (7) and Proposition 1.

□

PROPOSITION 3. The solution to problem (1) can be represented by

$$u(x) = \lim_{t \uparrow \infty} \left\{ E^x\ [\tau \leq t,\ e_{\ q}(\tau)\, f(x_t)] - \sum_{\lambda_k > 0} \frac{\mu - \lambda_k}{\lambda_k}\ (u^{\mu}, \varphi_k) e^{\lambda_k t} \varphi_k(x) \right\}\ .$$

PROOF. Set $u^+ = \sum_{\lambda_k > 0} (u, \varphi_k)\varphi_k$ and

$u^- = \sum_{\lambda_k<0} (u,\varphi_k)\varphi_k$. It is easy to see that

$$\lim_{t\uparrow\infty} T_t u^-(x) = 0 , \qquad \forall\ x \in D \tag{8}$$

and

$$T_t u^+ = \sum_{\lambda_k>0} (u,\varphi_k)\, e^{\lambda_k t}\, \varphi_k .$$

Then by Propositions 2 and (8), we have

$$u(x) = \lim_{t\uparrow\infty} \left\{ E^x \left[\tau \leqslant t,\ e_q(\tau)\, f(x_\tau)\right] + \sum_{\lambda_k>0} (u,\ \varphi_k)\, e^{\lambda_k t} \varphi_k(x) \right\} . \tag{9}$$

By Proposition 1, $\forall\ k \geqslant 1$,

$$(u,\varphi_k) = (u^\mu,\varphi_k) + \mu \frac{\int_0^t e^{\lambda_k s}\, ds}{1 - e^{\lambda_k t}} (u^\mu,\ \varphi_k)$$

$$= \frac{\lambda_k - \mu}{\lambda_k} (u^\mu,\ \varphi_k) . \tag{10}$$

Thus, Proposition 3 follows from (9) and (10).

□

Remark. Since $\forall\ t > 0$,

$$E^x \left[e_q(\tau \wedge t)\ u^\mu\ (x_{\tau\wedge t})\right] = E^x \left[\tau \leqslant t,\ e_q(\tau)\ f(x_\tau)\right]$$

$$+\ E^x \left[t < \tau,\ e_q(t)\ u^\mu(x_t)\right]$$

$$=\ E^x \left[\tau \leqslant t,\ e_q(\tau)\, f(x_\tau)\right] + T_t\, u^\mu(x) ,$$

We obtain by Proposition 3,

$$u(x) = \lim_{t\uparrow\infty} \left\{ E^x \left[e_q(\tau\wedge t)\, u^\mu(x_{\tau\wedge t})\right] - \sum_{\lambda_k>0} \frac{\mu}{\lambda_k} (u^\mu,\varphi_k)\, e^{\lambda_k t} \varphi_k(x) \right\}.$$

This is an equivalent form given by Zhiming Ma.

References

[1] K. L. CHUNG and ZHONGXIN ZHAO. From Brownian Motion to Schrödinger Equation, to appear.

[2] B. SIMON. Schrödinger Semigroups. Bull. Amer. Math. Soc. 7 (1982), 447-526.

Zhiming Ma
Institute of Applied Mathematics
Academia Sinica
Beijing, China

Zhongxin Zhao
Institute of Systems Science
Academia Sinica
Beijing, China

SUBORDINATORS REGENERATED*

by

B. MAISONNEUVE

1. Introduction

Subordinators are much older objects than regenerative sets, as defined by Hoffmann - Jorgensen [3], and it is natural to try to deduce the structure of the second from the structure of the first. This was done in [5] through the *existence* of a local time for a perfect regenerative set.

Here we shall show that regenerative sets can do much for subordinators as well. In fact, Lévy's decomposition ant Lévy-Khinchin formula can be immediately derived from the fact that the image of a subordinator is a regenerative set (see Meyer [7]) and from an easy result of *uniqueness* for the continuous additive functionals with support in a regenerative set.

2. Basic definitions for regenerative sets

Let $(\Omega, \underline{\underline{F}}, P)$ be a complete probability space and let $(\underline{\underline{F}}_t)$ be a filtration of $\underline{\underline{F}}$, with the usual conditions. The set Ω is furnished with a semi-group of shifts $(\theta_t)_{t \in \mathbf{R}_+}$ such that the mapping $(t, \omega) \to \theta_t \omega$ is $(\mathbf{R}_+ \times \underline{\underline{F}}, \underline{\underline{F}})$ measurable. We set $\theta_\infty \omega = \omega_\Delta$, where ω_Δ is a distinguished point of Ω.

(2.1) DEFINITION. — *Let M be a right closed progressive random set. The collection $(\Omega, \underline{\underline{F}}, \underline{\underline{F}}_t, M, \theta_t, P)$ is said to be a regenerative set provided that*

$$M \circ \theta_t = (M - t) \cap \mathbf{R}_+ \ , \quad t \in \mathbf{R}_+ \ ,$$

and that the following regeneration property holds : for every stopping time T

* This work was supported by AFOSR contract N° F49620-79-C-0080 while the author was visiting Northwestern University, Evanston, in August 1979.

such that $T \in M$ a.s. *on* $\{T < \infty\}$, *and every bounded* $\underset{=}{F}$*-measurable function* f ,

$$E[f \circ \theta_T \,|\underset{=T}{F}] = E[f] \quad \text{a.s. on} \quad \{T < \infty\} . \tag{2.2}$$

This is essentially the definition of Hoffmann - Jorgensen [3]. We associate with M the following random variables :

$$D_t = \inf\{s > t \ : \ s \in M\} \ , \quad t \in \mathbf{R}_+ \ .$$

3. Continuous additive functionals with support in $\overline{M}$

Let $(\Omega, \underset{=}{F}, \underset{=t}{F}, M, \theta_t, P)$ be a fixed regenerative set such that M has a.s. *no isolated points.* Under this assumption we know from [4], [5] or [6] that there exists a perfect continuous additive functional whose support is $\overline{M}$. Such an additive functional is called a *local time.* The additivity refers to the shifts (θ_t) ; by the support of an increasing process B we mean the support of the associated random measure, or the set of all points of increase of B . It will be denoted by supp$B$. Here is our main result.

(3.1) THEOREM. — *Let* L *be a local time and let* B *be a continuous additive functional such that* supp$B \subset \overline{M}$ a.s. *Then,*

$$u_B^1 \equiv E\left[\int_0^\infty e^{-s} dB_s\right] < +\infty \ ; \tag{3.2}$$

in particular, $u_L^1 < +\infty$.

$$B = aL \quad \text{a.s., where} \quad a = u_B^1 / u_L^1 \ \ (0/0 = 0) \ . \tag{3.3}$$

(3.4) REMARK. — This result is analogous to Theorem (3.13) of Blumenthal, Getoor [1], p. 216. If you think it is wrong, read § 6.

Proof of Theorem (3.1). —

1) Set $T_t = \inf\{s : B_s > t\}$. One has

$$T_{t+s} = T_t + T_s \circ \theta_{T_t} \quad \text{a.s. on} \quad \{T_t < \infty\}$$

and therefore

$$\int_0^\infty e^{-s} dB_s = \sum_{n=0}^\infty e^{-T_n} \left(\int_0^{T_1} e^{-s} dB_s\right) \circ \theta_{T_n} \leq \sum_{n=0}^\infty e^{-T_n} \ .$$

We can apply the regeneration property at each T_n to get

$$u_B^1 \leq \sum_{n=0}^\infty \left(E\left[e^{-T_1}\right]\right)^n < +\infty \ .$$

2) The assumption that supp$B \subset \overline{M}$ and the regeneration property at time D_t imply

$$E\left[\int_t^\infty e^{-s} dB_s | \underset{=D_t}{F}\right] = E\left[\int_{D_t}^\infty e^{-s} dB_s | \underset{=D_t}{F}\right] = u_B^1 e^{-D_t} .$$

This applies to L as well, and it follows that

$$E\left[\int_t^\infty e^{-s} dB_s | \underset{=D_t}{F}\right] = E\left[\int_t^\infty e^{-s} a dL_s | \underset{=D_t}{F}\right] ,$$

which implies that $B = aL$ a.s.

4. Lévy's decomposition of a subordinator

Let (S_t) be a subordinator with values in $\overline{\mathbf{R}}_+$, such that $S_0 = 0$, defined on some stochastic base $(\Omega, \underset{=}{G}, \underset{=t}{G}, P)$ with the usual conditions. For every $(\underset{=t}{G})$ stopping time T and every bounded path function f one has

(4.1) $$E[f(S_{T+.} - S_T) | \underset{=T}{G}] = E[f(S.)] \quad \text{a.s. on} \quad \{S_T < \infty\} .$$

(4.2) THEOREM (Lévy's decomposition). — *There exists $a \in \mathbf{R}_+$ such that a.s.*

(4.3) $$S_t = at + \sum_{0<s\le t} \Delta S_s , \quad t \in \mathbf{R}_+ .$$

Proof. — Let $M = \{t \in \mathbf{R}_+ : \ S_s = t \ \text{ for some } \ s\}$. If the first point of increase of S is a.s. greater than 0 , then the process S is a pure jump process, and (4.3) holds with $a = 0$. Let us now assume that the first point of increase of S is a.s. 0 . Then every $t \in \mathbf{R}_+$ is a point of right increase a.s. on $\{S_t < \infty\}$, and the process S is a.s. strictly increasing until it reaches $+\infty$. It is then convenient to assume that Ω is just the set of all right continuous functions ω from $\mathbf{R}_+$ into $\overline{\mathbf{R}}_+$, such that $\omega(0) = 0$, which are strictly increasing until they reach $+\infty$, and to assume that (S_t) is the process of the coordinates of Ω .

Let (L_t) be the right continuous inverse of (S_t) . With $\underset{=t}{F} = \underset{=L_t}{G}$ and $\theta_t = S_{L_t+.} - t$, the collection

$$(\Omega, \underset{=}{G}, \underset{=t}{F}, M, \theta_t, P)$$

is a *regenerative set*, as it is easily checked (see [2] for a more general result), and *L is a local time* of this regenerative set; in fact L is continuous, additive, and its support is $\overline{M}$.

Now let us consider the continuous additive functional

$$B_t = Leb(M \cap [0,t]) .$$

Its support is contained in $\overline{M}$. Therefore, $B = aL$ a.s. for some $a \in \mathbf{R}_+$, by Theorem (3.1). One has identically

$$S_t = B_{S_t} + \sum_{0<s\leq t} \Delta S_s .$$

But $B_{S_t} = aL_{S_t} = at$ a.s. and the proof is complete.

(4.4) REMARK. — The difficult part of the theory of local times for regenerative sets is the existence, but here L is already constructed as the inverse of S .

5. Lévy-Khinchin formula

Let $(\Omega, \underline{\underline{G}}, \underline{\underline{G}}_t, S_t, P)$ be a subordinator. We define M as in § 4, and let L be the right continuous inverse of S . It follows from § 4 that there exists a unique $a \in \mathbf{R}_+$ such that a.s.

(5.1) $$Leb(M \cap [0,t]) = aL_t \ , \ \ t \in \mathbf{R}_+ .$$

Of course, if M is a.s. discrete, one has $a = 0$.

(5.2) THEOREM (Lévy-Khinchin). — *For every positive Borel function h* on R^+ which vanishes at 0

(5.3) $$\lambda(h) = E\left[\sum_{0<s\leq 1} h(\Delta S_s)\right] .$$

Then for every $p \in (0,\infty)$ one has $(e_p(x) = e^{-px})$

(5.4) $$\log E[e_p(S_t)] = -t[pa + \lambda(1-e_p)] \ , \ \ t \in \mathbf{R}_+ ,$$

provided that the subordinator (S_t) assumes only finite values. In particular $\lambda(1-e_p) < \infty$ for every $p \in (0,\infty)$.

Let us first prove a lemma which tells us that (t, λ) is the "Lévy system" of S .

(5.5) LEMMA. — *If $S_t < \infty$ for every t one has*

(5.6) $$E[\sum Z_s h(\Delta S_s)] = \lambda(h) E\left[\int_0^\infty Z_s ds\right]$$

for every positive $(\underline{\underline{G}}_t)$ predictable Z and every positive Borel function h on $\mathbf{R}_+$ which vanishes at 0 .

Proof. — By homogeneity,

$$E\left[\sum_{0<s\leq t} h(\Delta S_s)\right] = t\lambda(h) \ , \ t \in \mathbf{R}_+ .$$

Therefore, (5.6) holds whenever $\lambda(h)$ is finite, for instance if $h = 1_{(c,\infty)}$ for $c > 0$ or if h is bounded by K and carried by (c, ∞) . Letting $K \nearrow \infty$ and $c \searrow 0$, we get (5.6) for all positive h .

Proof of theorem (5.2). — One has the following obvious identity :

$$1 - e_p(S_t) = p \int_0^{S_t} e_p(s)ds = p \int_0^{S_t} e_p(s) I_{\{s \in M\}} ds + \sum_{0 < s \leq t} e_p(S_{s-})(1 - e_p(\Delta S_s)) \ .$$

But,

$$\int_0^{S_t} e_p(s) I_{\{s \in M\}} ds = a \int_0^{S_t} e_p(s) dL_s = a \int_0^t e_p(S_s) ds \ .$$

Therefore, taking the expectation in (5.7) and applying (5.6) with $Z_s = e_p(S_{s-}) 1_{s \leq t}$ and $h = 1 - e_p$, we get

$$1 - E[e_p(S_t)] = [ap + \lambda(1 - e_p)] \int_0^t E[e_p(S_s)] ds \ , \ t \in \mathbf{R}_+ \ .$$

This shows that $\lambda(1 - e_p) < \infty$ ($E[e_p(S_s)] > 0$ for all s) and that the map $t \to E[e_p(S_t)]$ is continuous and satisfies (5.4).

(5.8) REMARKS. —

a) If the subordinator can assume the value $+\infty$ there still exits a measure λ for which (5.4) holds, but it puts mass on $\{+\infty\}$ and does not satisfy (5.3).

b) (5.4) follows also from Ito's decomposition and from the fact that the jumps of S define a Poisson random measure. This is a classical derivation.

6. Some complements on regenerative sets

Let us first make some comments on definition (2.1) and theorem (3.1). If we weaken the regeneration property by requiring (2.2) only for $f \in b\, \underline{\underline{F}}^M$, where $\underline{\underline{F}}^M$ is the smallest σ-field on Ω that makes the random set M measurable, then the statement of theorem (3.1) becomes *wrong*. The following example is due to Çinlar. Let $(\Omega^i, \underline{\underline{F}}^i, \underline{\underline{F}}^i_t, M^i, \theta^i_t, P^i)$, $i = 1, 2$ be two regenerative sets in the sense of definition (2.1) and set $\Omega = \Omega^1 \times \Omega^2$, $\underline{\underline{F}} = \underline{\underline{F}}^1 \times \underline{\underline{F}}^2$, $\underline{\underline{F}}_t = \underline{\underline{F}}^1_t \times \underline{\underline{F}}^2_t$, $M^i(\omega^1, \omega^2) = M^i(\omega^i)$, $i = 1, 2$, $\theta_t(\omega^1, \omega^2) = (\theta^1_t \omega^1, \theta^2_t \omega^2)$, $P = P^1 \times P^2$. Then over $(\Omega, \underline{\underline{F}}, \underline{\underline{F}}_t, \theta_t, P)$, M^1 and M^2 are regenerative sets in the weak sense, not in the sense of definition (2.1). If the times $\inf\{t : t \notin M^i\}$, $i = 1, 2$ are exponential with parameters in $(0, \infty)$, then $M^1 \cap M^2 \neq \emptyset$ a.s., $\overline{M}^1 \neq \overline{M}^2$ a.s. and a local time B of the regenerative set $M^1 \cap M^2$ cannot be written in the form aL , by means of a local time L of M^1 and $a \in \mathbf{R}_+$.

Now let us note the following consequence of theorem (3.1), which also fails for weak regenerative sets.

(6.1) PROPOSITION. — *If M^1 and M^2 are two nonempty regenerative sets (in the sense of definition (2.1)), without isolated points, relative to the same base* $(\Omega, \underline{\underline{F}}, \underline{\underline{F}}_t, \theta_t, P)$ *then* $\overline{M}^1 = \overline{M}^2$ a.s. .

(6.2) REMARK. — Our assumptions imply that $0 \in M^1$ a.s. and $0 \in M^2$ a.s., so that $M^1 \cap M^2 \neq \emptyset$ a.s. . But in the case of possibly "delayed" regenerative sets the conclusion of proposition (6.1) should be that either $M^1 \cap M^2 = \emptyset$ a.s. or $\overline{M}^1 = \overline{M}^2$ a.s. . Theorem (3.1) extends without change to such delayed regenerative sets.

Proof . — Proposition (6.1) can easily be derived from theorem (3.1) and from the existence of local times for M^1 and M^2 . We prefer to give a direct argument. One has $0 \in M^2$ a.s. and, by the regeneration property for $M^1, D_t^1 \equiv \inf\{s > t : s \in M^1\} \in M^2$ a.s. on $\{D_t^1 < \infty\}$. Therefore $\bigcup_{t \text{ rational}} [D_t^1] \subset M^2$ a.s. and $\overline{M^1} \subset \overline{M^2}$ a.s. . Interchanging M^1 and M^2 , we have also $\overline{M^2} \subset \overline{M^1}$ a.s. and the proof is complete.

(6.3) FINAL REMARK. — Theorem (3.1) and proposition (6.1) still hold if we relax the homogeneity property of M in the following way :

$$M \circ \theta_t \cap]0, \infty[= (M - t) \cap]0, \infty[\ , \ t \in M \ .$$

Similarly the additivity of an additive functional B can be relaxed as follows :

$$B_{t+s} = B_t + B_s \circ \theta_t \ , \ t \in M \ , \ s \in \mathbf{R}_+ \ .$$

References

[1] R.M. BLUMENTHAL AND R.K. GETOOR. — *Markov Processes and Potential Theory*, Academic Press, New-York, 1968.

[2] E. ÇINLAR AND B. MAISONNEUVE. — The image of a markov additive process, (to appear).

[3] J. HOFFMANN-JORGENSEN. — *Markov sets*, Mathematica Scandinavia, 24, fasc. 2,(1969).

[4] B. MAISONNEUVE AND PH. MORANDO. — *Temps local d'un ensemble régénératif*, Séminaire de Probabilités IV. Lecture Notes in Mathematics, Springer, Berlin, 124, 1970.

[5] B. MAISONNEUVE. — *Ensembles régénératifs, temps locaux et subordinateurs*, Séminaire de Probabilités V. Lecture Notes in Mathematics, Springer, Berlin, 191, 1971.

[6] B. MAISONNEUVE. — *Systèmes Régénératifs*, Astérisque 15, S.M.F., 1974.

[7] P.A. MEYER. — *Ensembles régénératifs, d'après Hoffmann- Jorgensen*, Séminaire de Probabilités IV. Lecture Notes in Mathematics, Springer, Berlin, 124, 1970.

POSTSCRIPT. —

This paper was written in August 1979 and submitted to - but never received by - the russian journal Theory of Probability and its Applications. I think that the paper is not yet obsolete, since (3.1) provides (77) of

J. AZEMA. — *Sur les fermés aléatoires*. Séminaire de Probabilités XIX. Lecture Notes in Mathematics, 1123, Springer, Berlin, (1985).

under weaker assumptions and with a simpler proof. For a more detailed discussion concerning weak regeneration (as defined in section 6) and intersections of regenerative sets, the reader is referred to :

P.J. FITZSIMMONS, B. FRISTEDT and B. MAISONNEUVE. — *Intersections and limits of regeneratrice sets*. Z. Wahrscheinlichkeits theorie verw. Gebiete. **70**, (1985), 157-173.

B. MAISONNEUVE
I.M.S.S.
Universite de Grenoble II
47X-38040 Grenoble Cedex
FRANCE

LOCAL NONDETERMINISM AND HAUSDORFF DIMENSION[1]

by

Ditlev Monrad and Loren D. Pitt

§1. INTRODUCTION

Let $X = \{X(t,\omega): t \in \mathbb{R}^N\}$ denote d-dimensional fractional Brownian motion of index $\alpha \in (0,2)$, normalized such that $X(0) = 0$ and

$$E[e^{i\langle u,X(t)-X(s)\rangle}] = e^{-\frac{1}{2}|u|^2|t-s|^\alpha}$$

for all $u \in \mathbb{R}^d$ and $s,t \in \mathbb{R}^N$. If $\alpha = 1$, then we have Lévy's multiparameter Brownian motion.

The process X is locally nondeterministic [28], and we use this to show that if $2N \leq \alpha d$, then

$$\dim X(E) = 2/\alpha \dim(E)$$

for every closed set $E \subseteq \mathbb{R}^N$, almost surely. In other words the exceptional null set does not depend on E. It

[1]Supported in part by the National Science Foundation.

is this uniform nature of our results which distinguishes this result from those of Kahane in [18,19].

The case $\alpha d < 2N$ is quite different. In this case X has a continuous local time. For each fixed, nonrandom, closed set $E \subset \mathbb{R}^N$,

$$(*) \qquad \dim X(E) = \min\{d, \tfrac{2}{\alpha} \dim(E)\},$$

almost surely, see [18]. But here the exceptional null set does depend on E. In fact, we will show that

$$\dim X^{-1}(F) = N - \tfrac{1}{2}\alpha d + \tfrac{1}{2}\alpha \dim(F)$$

for every closed set $F \subseteq R^d$, almost surely, which shows the exceptional set for (*) must depend on E. This uniform result thus strengthens the results of Testard in [31]. These results extend to a large class of locally nondeterministic Gaussian random fields.

There is a large literature on nonuniform results which begins with Taylor's 1955 paper [32] on the level sets of Brownian motion. For more details we refer to the bibliography and the books of Adler [3] and Kahane [19].

Uniform dimension results such as ours have previously been proved by Kaufman [20], [21], and [22] in the case of Brownian motion. Uniform dimension results for the inverse images $X^{-1}(\{x\})$ of singletons were obtained by Berman [5] for a large class of locally nondeterministic real valued Gaussian processes $\{X(t):t \geq 0\}$ with stationary increments, and for the Brownian sheet by Adler [2]. Hawkes and Pruitt [17] got uniform dimension results for

$X(E)$ in the case when $\{X(t): t \geq 0\}$ is a strictly stable Lévy process of index α with $\alpha \leq d$.

The precise statement of our results is given in Section 2. The proofs are given in Sections 3, 4, and 5. Finally, we have included an Appendix on the spectral analysis of stationary Gaussian random fields.

§2. DEFINITIONS AND STATEMENT OF RESULTS

Let $X = \{X(t): t \in \mathbb{R}^N\}$ be a real valued, centered, Gaussian stochastic process. Let $0 < \alpha < 2$. We shall say that X is <u>α-nondeterministic</u> on an open set $G \subset \mathbb{R}^N$ if there exist two strictly increasing continuous functions

$$\varphi, \psi: [0,\infty) \longrightarrow [0,\infty)$$

such that

(2.1) $\varphi(0) = \psi(0) = 0$. $\varphi(h) \leq \psi(h)$ for all $h \geq 0$.

(2.2) $\lim_{h\downarrow 0} \psi(h)/h^{\beta} = 0$ for every $\beta < \alpha$.

(2.3) $\lim_{h\downarrow 0} h^{\beta}/\varphi(h) = 0$ for every $\beta > \alpha$.

(2.4) $\mathrm{Var}(X(s)-X(t)) \leq \psi(|s-t|)$ for $s,t \in G$.

(2.5) For each integer $n \geq 2$ there exists a constant $\varepsilon_n > 0$ such that whenever $t_1,\ldots,t_n \in G$ satisfy

$$|t_n-t_{n-1}| \leq |t_n-t_i| \qquad i = 1,\ldots,n-1$$

we have

$$\mathrm{Var}\{X(t_n)|X(t_1),\ldots,X(t_{n-1})\} \geq \varepsilon_n \varphi(|t_n-t_{n-1}|).$$

If we can choose $\varepsilon_n = \varepsilon$ for all n in (2.5), then we shall say that X is <u>strongly α-nondeterministic</u> on G.

We shall say that X is <u>locally (strongly) α-nondeterministic</u> on G if X is (strongly) α-nondeterministic on a neighborhood of every point in G.

REMARK. Condition (2.5) is equivalent to

(2.6) For each $n \geq 2$ there exists $c_n > 0$ such that for arbitrary $u_1, \ldots, u_n \in R$ and any $t_1, \ldots, t_n \in G$ satisfying

$$|t_{j+1}-t_j| \leq |t_{j+1}-t_i| \text{ for } 1 \leq i \leq j < n$$

we have

$$\mathrm{Var}[u_1 X(t_1) + \sum_2^n u_j(X(t_j)-X(t_{j-1}))]$$

$$\geq c_n[u_1^2 + \sum_2^n u_j^2 \varphi(|t_j-t_{j-1}|)].$$

The original definition of local nondeterminism was given by Berman [6] with $N = 1$ and Pitt [28] for $N > 1$. Further refinements and applications of these ideas may be found in Berman [8] in Cuzick [9,10], Cuzick and du Preez [12] and Nolan [26].

Recall the definition of Hausdorff dimension. For any subset S of $\mathbb{R}^k$ and each $p \geq 0$, $\delta > 0$ we define

$$\mu_\delta^p(S) = \inf \sum_i [d(B_i)]^p,$$

where the infimum is taken over all covers of S by a collection of balls $\{B_i\}$ of diameters $d(B_i) < \delta$. The p-dimensional outer Hausdorff measure of S is

$$\mu^p(S) = \lim_{\delta \downarrow 0} \mu^p_\delta(S).$$

The Hausdorff dimension of S is defined

$$\dim(S) = \inf\{p: \mu^p(S) = 0\}$$
$$= \sup\{p: \mu^p(S) = \infty\}.$$

An excellent general reference on Hausdorff measures and dimension is [29].

We are now able to state our results. The results of Pitt in [28] on the local times of locally nondeterministic Gaussian random fields are easily modified to give us results on the existence of continuous local times for a broad class of locally α-nondeterministic random fields, and this class includes the fractional Brownian motions.

THEOREM 1. Let $X = \{(X_1(t),\ldots,X_d(t)): t \in \mathbb{R}^N\}$ be a d-dimensional centered Gaussian random field with continuous sample functions. Assume that the processes $X_1,\ldots,X_d$ are independent and that for some α with $0 < \alpha < 2$ each is locally α-nondeterministic. Assume that $\alpha d < 2N$. For each rectangle $[s,t] \subset \mathbb{R}^N$, let $\varphi([s,t],x)$ denote the sojourn density of $\{X(\tau): \tau \in [s,t]\}$, chosen so that $(s,t,x) \longrightarrow \varphi([s,t],x)$ is jointly continuous. Consider the open set

$$(2.7) \qquad O = \bigcup_{[s,t]} \{x \in \mathbb{R}^d: \varphi([s,t],x) > 0\}.$$

With probability one,

$$\dim X^{-1}(F) = N - \frac{1}{2}\alpha d + \frac{1}{2}\alpha \dim(F)$$

for every closed set $F \subset O$. (The exceptional null set does not depend on F.)

EXAMPLE 1. For fractional Brownian motion it follows from the scaling property that $O = \mathbb{R}^d$, almost surely (in the case $\alpha d < 2N$).

The following example shows that O can be a proper subset of $\mathbb{R}^d$.

EXAMPLE 2. Consider the real valued stationary Gaussian process $X = \{X(t): -\infty < t < \infty\}$ defined by

$$X(t) = \frac{\sqrt{8}}{\pi} \sum_{n=1}^{\infty} \frac{1}{(2n-1)} [\xi_n \cos((2n-1)t) + \eta_n \sin((2n-1)t)],$$

where $\{\xi_n\}$ and $\{\eta_n\}$ are independent, normal (0,1).

$$E[X(0)X(t)] = 1 - \frac{2}{\pi} |t| \text{ for } -\pi \le t \le \pi.$$

It follows (see Berman [7]) that X is locally strongly α-nondeterministic with $\alpha = 1$. The fact that the stationary process in the above example has a discrete spectral measure is essential, as the following theorem shows.

THEOREM 2. Let $\{X(t) \in \mathbb{R}^d: t \in \mathbb{R}^N\}$ be a measurable stochastically continuous stationary Gaussian random field. Assume that for each rectangle $I \subset \mathbb{R}^N$ the sojourn density $\varphi(I,x)$ is continuous and that for $n = 1,2,...$

$$E[\max_{|x| \le n} \varphi(I,x)] < \infty. \tag{2.8}$$

Let $\Delta_c(d\lambda)$ denote the continuous part of the spectral measure of X and assume that the covariance matrix

$$\Sigma = \Delta_c(\mathbb{R}^N)$$

is nonsingular. Then, almost surely,

$$\bigcup_I \{x \in \mathbb{R}^d: \varphi(I,x) > 0\} = \mathbb{R}^d. \tag{2.9}$$

REMARK. Any stationary Gaussian random field satisfying the assumptions of Theorem 1 will automatically satisfy (2.8). See e.g., the proof of Theorem (27.1) of [14].

We now turn to uniform dimension results for X(E).

THEOREM 3. Let $X = \{(X_1(t),...,X_d(t)): t \in \mathbb{R}^N\}$ be a d-dimensional centered Gaussian random field with continuous sample functions. Assume that the processes $X_1,...,X_d$ are independent and that for some $\alpha \in (0,2)$ each is locally strongly α-nondeterministic. Assume that $2N \le \alpha d$. With probability one,

$$\dim X(E) = \frac{2}{\alpha} \dim(E)$$

for every closed set $E \subset \mathbb{R}^N$.

REMARK. In the one-parameter case ($N = 1$), Theorem 3 still holds true if we replace the 2-sided assumption

$$\text{(2.10)} \qquad \operatorname{Var}\{X(t) \mid X(s): h \le |s-t| \le \varepsilon\} \ge K \cdot \varphi(h)$$

by the weaker 1-sided assumption

$$\text{(2.11)} \qquad \operatorname{Var}\{X(t+h) \mid X(s): t-\varepsilon \le s \le t\} \ge K \cdot \varphi(h)$$

§3. PROOF OF THEOREM 1.

We begin with a real variable lemma. We shall use the notation $\underline{t} = (t_1, \ldots, t_N)$ for points in $\mathbb{R}^N$ to distinguish between the points in $\mathbb{R}^N$ and their coordinates. As usual, $0 < \alpha < 2$.

LEMMA 3.1. Assume that $\{\underline{f}(\underline{t}) \in \mathbb{R}^d : \underline{t} \in [0,1]^N\}$ satisfies a uniform Hölder condition of every order smaller than $\frac{1}{2}\alpha$, and that $\underline{f}$ has a bounded sojourn density $\{\varphi(y): y \in \mathbb{R}^d\}$. Then

$$\text{(3.1)} \qquad \frac{2}{\alpha} \dim(E) + d - \frac{2N}{\alpha} \le \dim \underline{f}(E) \le \frac{2}{\alpha} \dim(E)$$

for every closed set $E \subset [0,1]^N$.

PROOF. The upper bound is well known. To prove the lower bound we need only show that for any compact set $F \subset \mathbb{R}^d$,

(3.2) $$\dim \underline{f}^{-1}(F) \leq N - \frac{1}{2}\alpha d + \frac{1}{2}\alpha \dim(F).$$

For $n = 1,2,3,\ldots$ and $\underline{k} = (k_1,\ldots,k_N) \in \mathbb{R}^N$, where each $k_i \in \{1,\ldots,2^n\}$, let

(3.3)

$$I_{n,\underline{k}} = \{\underline{t} \in [0,1]^N: (k_i-1)2^{-n} \leq t_i \leq k_i 2^{-n} \text{ for all } i\}.$$

For $x \in \mathbb{R}^d$ and $\varepsilon > 0$, let

(3.4) $$B(x,\varepsilon) = \{y \in \mathbb{R}^d: |x-y| < \varepsilon\}.$$

Fix $\beta < \alpha$. There exists a constant $\varepsilon_0 > 0$ such that for any $x \in \mathbb{R}^d$ and any $\varepsilon < \varepsilon_0$, the following two conditions

(3.5) $$2^{-n\beta/2} \leq \varepsilon$$

(3.6) $$I_{n,\underline{k}} \cap \underline{f}^{-1}(B(x,\varepsilon)) \neq \phi$$

imply

(3.7) $$\underline{f}(I_{n,\underline{k}}) \subset B(x,2\varepsilon).$$

Now consider the number of cubes $I_{n,\underline{k}}$ satisfying (3.7),

(3.8) $$N(n,x,\varepsilon) = \#\{\underline{k}: \underline{f}(I_{n,\underline{k}}) \subset B(x,2\varepsilon)\}.$$

Since

$$N(n,x,\varepsilon)\cdot 2^{-nN} \leq \int_{B(x,2\varepsilon)} \varphi(y)dy \leq C'\varepsilon^d,$$

for some constant C' independent of n, x, and ε, we can conclude that

(3.9) $$N(n,x,\varepsilon) \leq C'\cdot 2^{nN}\cdot\varepsilon^d.$$

Let us return to the compact set $F \subset \mathbb{R}^d$. Choose η and p such that

(3.10) $$\dim F < \eta = \frac{2}{\beta} p - \frac{2}{\beta} N + d.$$

Consider a covering $F \subset \bigcup B(x_i,\varepsilon_i)$, for which $\varepsilon_i < \varepsilon_0$ for all i and

(3.11) $$\sum \varepsilon_i^{\eta}$$

is small. For each i choose n_i such that

$$2^{-n_i\beta/2} \le \varepsilon_i < 2^{-(n_i-1)\beta/2}.$$

It follows from (3.9) that

(3.12) $$N(n_i,x_i,\varepsilon_i) \le C\cdot 2^{n_i(N-\beta d/2)}.$$

For each i, let

(3.13) $$\mathcal{C}_i = \{I_{n_i,\underline{k}}: I_{n_i,\underline{k}} \cap \underline{f}^{-1}(B(x_i,\varepsilon_i)) \ne \phi\}.$$

Clearly,

$$\underline{f}^{-1}(F) \subset \bigcup_i \underline{f}^{-1}(B(x_i,\varepsilon_i)) \subset \bigcup_i \bigcup_{I\in\mathcal{C}_i} I.$$

It follows that

$$\sum_i \sum_{I\in\mathcal{C}_i} (\operatorname{diam} I)^p \le N^{p/2} \sum_i N(n_i,x_i,\varepsilon_i)\cdot 2^{-n_i p}$$

$$\le C\cdot N^{p/2} \sum 2^{-n_i(p-N+\beta d/2)}$$

$$\le C\cdot N^{p/2} \sum \varepsilon_i^{2p/\beta-2N/\beta+d} = C\cdot N^{p/2} \sum \varepsilon_i^{\eta}.$$

Since C only depends on $\underline{f}$, and (3.11) can be made arbitrarily small, this shows that

$$\dim \underline{f}^{-1}(F) \leq p = N - \frac{1}{2}\beta d + \frac{1}{2}\beta\eta. \tag{3.14}$$

This completes the proof. ■

We now turn our attention to a Gaussian random field $X = \{X(t) \in \mathbb{R}^d: t \in \mathbb{R}^N\}$ satisfying the assumptions of Theorem 1. Consider a fixed cube $I \subset \mathbb{R}^N$. We shall prove that with probability one,

$$\dim X^{-1}(F) \geq N - \frac{1}{2}\alpha d + \frac{1}{2}\alpha \dim(F) \tag{3.15}$$

for every closed set $F \subset \{x \in \mathbb{R}^d: \varphi(I,x) > 0\}$. For $\bar{t} = (t_1,\ldots,t_k) \in \mathbb{R}^{Nk}$ and $\bar{u} = (u_1,\ldots,u_k) \in \mathbb{R}^{dk}$, let

$$\hat{p}_k(\bar{t},\bar{u}) = \exp\{-\frac{1}{2}\operatorname{Var}\sum_{j=1}^{k}\langle u_j, X(t_j)\rangle\}.$$

For $0 < \gamma < 1$, k even, and any cube $J \subset I$ of sufficiently small diameter,

$$\begin{aligned} &E[(\varphi(J,x) - \varphi(J,y))^k] \\ &= (2\pi)^{-kd}\int_{J^k} d\bar{t}\int_{R^{dk}} \hat{p}_k(\bar{t},\bar{u}) \prod_{j=1}^{k} [e^{-i\langle u_j,x\rangle} - e^{-i\langle u_j,y\rangle}]d\bar{u} \\ &\leq C_\gamma^k \cdot |x-y|^{k\gamma} \cdot M_{k,\gamma}(J), \end{aligned} \tag{3.16}$$

where

$$M_{k,\gamma}(J) = \int_{J^k} d\bar{t}\int_{R^{dk}} \hat{p}_k(\bar{t},\bar{u}) \prod_{j=1}^{k} |u_j|^\gamma d\bar{u}. \tag{3.17}$$

Assume that $(\frac{1}{2}+\gamma)(\alpha+\gamma)d < N$. Using (2.3) and (2.6), we can easily modify the computations of sections 6 and 7 of [28] to prove that if J has sufficiently small diameter and $\overline{t} \in J^k$, then

$$\int_{R^{dk}} \hat{p}_k(\overline{t},\overline{u}) \prod_{j=1}^{k} |u_j|^\gamma d\overline{u} \le C_{k,\gamma} \prod_{j=2}^{k} |t_j - t_{j-1}|^{-(\frac{1}{2}+\gamma)(\alpha+\gamma)d}.$$

Hence

$$(3.18)\qquad M_{k,\gamma}(J) \le C^1_{k,\gamma}\cdot(\text{diam } J)^{k(N-(\frac{1}{2}+\gamma)(\alpha+\gamma)d)}.$$

We can now argue as in the proof of Theorem (27.1) of [14] to get

LEMMA 3.2. For any $\varepsilon > 0$ and any cube $J \subset I$ we have

$$(3.19)\qquad \max_x \varphi(J,x) \le C_\varepsilon(\omega)\cdot(\text{diam } J)^{N-\frac{1}{2}\alpha d-\varepsilon}.$$

Now consider a closed set $F \subset \{x \in \mathbb{R}^d : \varphi(I,x) > 0\}$. If $\dim(F) = 0$, let $\gamma = 0$. Otherwise, let $0 < \gamma < \dim(F)$. In either case, let μ be a positive measure on F satisfying

$$(3.20)\qquad \mu(S) \le (\text{diam } S)^\gamma.$$

for every set $S \subset \mathbb{R}^d$. Define the random measure λ on I by

$$(3.21)\qquad \lambda(B) = \int \varphi(B,x)\mu(dx)$$

for every Borel set $B \subset I$. Here we use the fact that for a.a. ω, the set functions $J \to \varphi(J,x)$ extend simultaneously to Borel measures $\varphi(B,x)$ which are functions of x. The measure λ is supported by $X^{-1}(F)$ and $\lambda(I) > 0$. For any cube $J \subset I$ and any $\varepsilon > 0$,

$$\text{(3.22)} \qquad \operatorname{diam} X(J) \le \tilde{C}_\varepsilon(\omega)\cdot(\operatorname{diam} J)^{\frac{1}{2}\alpha-\varepsilon},$$

since the sample functions of X satisfy a uniform Hölder condition of every order smaller than $\frac{1}{2}\alpha$. Therefore,

$$\begin{aligned}\text{(3.23)} \qquad \lambda(J) &\le \max_x \varphi(J,x)\cdot\mu(X(J)) \\ &\le \max_x \varphi(J,x)\cdot(\operatorname{diam} X(J))^{\gamma} \\ &\le C_\varepsilon(\omega)\cdot(\operatorname{diam} J)^{N-\frac{1}{2}\alpha d-\varepsilon+(\frac{1}{2}\alpha-\varepsilon)\gamma}.\end{aligned}$$

By standard density arguments, it follows that the Hausdorff dimension of $X^{-1}(F)$ is at least

$$\text{(3.24)} \qquad N - \tfrac{1}{2}\alpha d + \tfrac{1}{2}\alpha\gamma - \varepsilon(1+\gamma).$$

This completes the proof of Theorem 1. ■

§4. PROOF OF THEOREM 2.

Let $X = \{X(t) \in \mathbb{R}^d\colon t \in \mathbb{R}^N\}$ be a stationary Gaussian random field satisfying the assumptions of Theorem 2. For any $t_0 \in \mathbb{R}^N$, let $T(t_0)$ denote the linear shift operator defined by

$$\begin{aligned}\text{(4.1)} \qquad & T(t_0)[g(X(t_1),\ldots,X(t_n))] \\ &= g(X(t_0+t_1),\ldots,X(t_0+t_n)).\end{aligned}$$

Let $\mathcal{F}(X)$ denote the σ-field generated by X. Consider the invariant σ-field

(4.2) $\mathcal{F}(X,t_0) = \{A \in \mathcal{F}(X): T(t_0)1_A = 1_A \text{ a.s.}\}$.

Fix a cube $I \subset \mathbb{R}^N$ and a vector $t_0 \in \mathbb{R}^N$. As $n \to \infty$, the sequence of continuous random functions

(4.3) $$S_n: x \to \frac{1}{n+1} \sum_{k=0}^{n} \varphi(I+kt_0,x),\ x \in \mathbb{R}^d,$$

converges almost surely uniformly on every bounded subset of $\mathbb{R}^d$ t0 (a continuous version of) the random function

(4.4) $$S_\infty: x \to E\{\varphi(I,x)|\mathcal{F}(X,t_0)\},\ x \in \mathbb{R}^d.$$

This follows from assumption (2.8) and Mourier's ergodic theorem for stationary sequences of Banach space valued random vectors. (See [23] or [25]). The proof of Theorem 2 will be complete once we find a $t_0 \in \mathbb{R}^N$ and a cube $I \subset \mathbb{R}^N$ such that S_∞ is almost surely strictly positive on $\mathbb{R}^d$.

In the Appendix we have collected some facts from the spectral theory of stationary Gaussian random fields. Let

(4.5) $$\Delta = \Delta_d+\Delta_c$$

be the decomposition of the spectral measure of X into discrete and continuous parts. Write

(4.6) $$X(t) = X_d(t)+X_c(t),$$

where X_d and X_c are independent stationary Gaussian random fields with spectral measures Δ_d and Δ_c, respectively.

Let $\mathcal{F}(X_d)$ denote the σ-field generated by X_d and put

(4.7) $\mathcal{F}(X_d, t_0) = \{A \in \mathcal{F}(X_d): T(t_0)1_A = 1_A \text{ a.s.}\}.$

Let Σ denote the covariance of $X_c(0)$, and

(4.8) $$f(x) = (2\pi)^{-d/2}|\Sigma|^{-1/2}\exp\{-\tfrac{1}{2}\langle x, \Sigma^{-1}x\rangle\}$$

the density of $X_c(0)$. For any cube $I \in \mathbb{R}^N$ and any $t_0 \in \mathbb{R}^N$, introduce the continuous random functions,

(4.9) $$\Phi_I(x) = \int_I f(x - X_d(t))dt, \quad x \in \mathbb{R}^d$$

(4.10) $$\Psi_{I,t_0}(x) = E\{\Phi_I(x)|\mathcal{F}(X_d, t_0)\}, \quad x \in \mathbb{R}^d.$$

LEMMA 4.1. With probability one,

(4.11) $$\Psi_{I,t_0}(x) > 0 \text{ for all } x \in \mathbb{R}^d.$$

PROOF. Define for $r > 0$,

(4.12) $$f_0(r) = \min\{f(x): |x| \le r\}.$$

f_0 is decreasing, continuous, and strictly positive. For all t and x,

(4.13) $$f(x - X_d(t)) \ge f_0(|x| + |X_d(t)|).$$

For $n = 1, 2, \ldots$, define the random variable Z_n by

(4.14) $$Z_n = \int_I f_0(n + |X_d(t)|)dt.$$

Since

$$E \int_I |X_d(t)|dt = \int_I E|X_d(t)|dt < \infty,$$

$Z_n > 0$, a.s. For all x with $|x| \le n$,

$$\Psi_{I,t_0}(x) = E\{\Phi_I(x)|\mathcal{F}(X_d,t_0)\} \tag{4.15}$$

$$\ge E\{Z_n|\mathcal{F}(X_d,t_0)\} > 0,$$

almost surely. This completes the proof. ∎

For bounded Borel sets $B \subset \mathbb{R}^N$ and arbitrary Borel sets $A \subset \mathbb{R}^d$ define the occupation measure

$$\mu(B,A) = \int_B 1_A(X(t))dt. \tag{4.16}$$

LEMMA 4.2. For any cube $I \subset \mathbb{R}^N$, any Borel set $A \subset \mathbb{R}^d$, and any $t_0 \in \mathbb{R}^N$, satisfying

$$\Delta_c\{\lambda \in \mathbb{R}^N: \langle t_0,\lambda\rangle = a\} = 0, \text{ for all } a, \tag{4.17}$$

we have

$$\lim_{n\to\infty} \frac{1}{n+1} \sum_{k=0}^{n} \mu(I+kt_0,A) = \int_A \Psi_{I,t_0}(x)dx \tag{4.18}$$

almost surely and in L^1.

REMARK. Since for any bounded Borel set A,

$$\frac{1}{n+1} \sum_{k=0}^{n} \mu(I+kt_0,A) = \int_A S_n(x)dx \to \int_A S_\infty(x)dx, \tag{4.19}$$

almost surely, it follows that if t_0 satisfies (4.17), then $S_\infty(x) = \Psi_{I,t_0}(x)$ for all x, almost surely. In view of Lemma 4.1, this completes the proof of Theorem 2.

PROOF OF LEMMA 4.2. Since

$$E[\mu(I,A)] = \int_I E[1_A(X(t))]dt < \infty,$$

it follows from the Birkhoff-Khinchin individual ergodic theorem that

$$(4.20) \qquad \frac{1}{n+1}\sum_{k=0}^{n} \mu(I+kt_0,A) \to E\{\mu(I,A)|\mathcal{F}(X,t_0)\},$$

almost surely. But by Proposition A.2. (in Appendix)

$$(4.21) \qquad \mathcal{F}(X,t_0) = \mathcal{F}(X_d,t_0) \subset \mathcal{F}(X_d).$$

Hence

$$(4.22) \quad E\{\mu(I,A)|\mathcal{F}(X,t_0)\} = E\{E\{\mu(I,A)|\mathcal{F}(X_d)\}|\mathcal{F}(X_d,t_0)\}.$$

Using the notation (4.8) and (4.9) we have

$$\begin{aligned}
E\{\mu(I,A)|\mathcal{F}(X_d)\} &= E\{\int_I 1_A(X_d(t)+X_c(t))dt|\mathcal{F}(X_d)\} \\
&= \int_I E\{1_A(X_d(t)+X_c(t))|\mathcal{F}(X_d)\}dt \\
&= \int_I \{\int_A f(x-X_d(t))dx\}dt \\
&= \int_A \{\int_I f(x-X_d(t))dt\}dx \\
&= \int_A \Phi_I(x)dx.
\end{aligned}$$

Hence

$$E\{\mu(I,A)\,|\,\mathcal{F}(X,t_0)\} = E\{\int_A \Phi_I(x)dx\,|\,\mathcal{F}(X_d,t_0)\}$$
$$= \int_A E\{\Phi_I(x)\,|\,\mathcal{F}(X_d,t_0)\}dx$$
$$= \int_A \Psi_{I,t_0}(x)dx.$$

This completes the proof. ■

§5. PROOF OF THEOREM 3.

We shall prove that with probability one,

$$\dim X(E) = \frac{2}{\alpha}\dim(E) \tag{5.1}$$

for every closed set $E \subset (0,1)^N$. Without loss of generality we may assume that each X_i is strongly α-nondeterministic on $(0,1)^N$.

The upper bound,

$$\dim X(E) \le \frac{2}{\alpha}\dim(E) \tag{5.2}$$

follows from the fact that the sample functions of X satisfy a uniform Hölder condition of every order smaller than $\frac{1}{2}\alpha$ on $(0,1)^N$. To prove the lower bound we need only show that for every compact $F \subset \mathbb{R}^d$,

$$\dim\{t \in (0,1)^N : X(t) \in F\} \le \frac{\alpha}{2}\dim(F). \tag{5.3}$$

We follow Kaufman ([20]). For $n = 1,2,\ldots$ and $\underline{k} = (k_1,\ldots,k_N) \in \mathbb{R}^N$, where each $k_i \in \{1,2,3,\ldots,4^n\}$, let

(5.4)

$$I_{n,\underline{k}} = \{\underline{t} \in [0,1]^N\colon (k_i-1)4^{-n} \le t_i \le k_i 4^{-n} \text{ for all } i\}.$$

The inequality (5.3) will follow from

LEMMA 5.1. Let $0 < \alpha-\varepsilon < \beta < \alpha$. For large enough n there exists no ball $B \subset \mathbb{R}^d$ of radius $2^{-n\beta}$ for which $X^{-1}(B)$ intersects more than $2^{n\varepsilon d}$ cubes $I_{n,\underline{k}}$.

In view of the fact that the sample functions of X satisfy a uniform Hölder condition of every order smaller than $\frac{1}{2}\alpha$, Lemma 5.1 follows from

LEMMA 5.2. For large enough n there does not exist more than $2^{n\varepsilon d}$ distinct points of the form $\underline{t}_j = 4^{-n}\underline{k}_j$, where $\underline{k}_j \in \{1,2,3,\ldots,4^n\}^N$, such that

$$(5.5) \qquad |X(\underline{t}_i)-X(\underline{t}_j)| < 3\cdot 2^{-n\beta} \quad \text{for } i \neq j.$$

PROOF. Let A_n denote the event that there does exist more than $2^{n\varepsilon d}$ such points. By an "A_n-tuple" we shall mean a set of n distinct points $\underline{t}_1,\ldots,\underline{t}_n$ of the form $\underline{t}_j = 4^{-n}\underline{k}_j$, where $\underline{k}_j \in \{1,2,\ldots,4^n\}^N$, such that

$$(5.6) \qquad |\underline{t}_{j+1}-\underline{t}_j| \le |\underline{t}_{j+1}-\underline{t}_i| \quad \text{for } i \le j \le n-1$$

in addition to

$$(5.7)\qquad |X(\underline{t}_{j+1})-X(\underline{t}_j)| < 3\cdot 2^{-n\beta} \quad \text{for } j = 1,\ldots,n-1.$$

(Any set of n distinct points in $[0,1]^N$ can be reordered so that they satisfy (5.6). It can be done in at least n different ways: Pick $\underline{t}_n$ first, then $\underline{t}_{n-1}$, then $\underline{t}_{n-2}$, etc.)

Let N_n denote the number of A_n-tuples. Let χ_1^2 denote the chi-square distribution with 1 degree of freedom. Choose $\gamma > \alpha$ such that $\gamma-\beta < \varepsilon$. For fixed $\underline{t}_1,\ldots,\underline{t}_n$ satisfying (5.6), we have by strong α-nondeterminism,

$$P\{|X(\underline{t}_{j+1})-X(\underline{t}_j)| < 3\ 2^{-n\beta} \quad \text{for } j = 1,\ldots,n-1\}$$

$$\le \prod_{i=1}^{d} P\{|X_i(\underline{t}_{j+1})-X_i(\underline{t}_j)| < 3\ 2^{-n\beta} \text{ for } j = 1,\ldots,n-1\}$$

$$\le [\prod_{j=1}^{n-1} P\{\chi_1^2 < C\ 4^{-n\beta)}|\underline{t}_{j+1}-\underline{t}_j|^{-\gamma}\}]^d$$

$$\le C_1^n\cdot 2^{n^2(\gamma-\beta)d} \prod_{j=1}^{n-1} |\underline{k}_{j+1}-\underline{k}_j|^{-\gamma d/2}.$$

It follows that

$$(5.8)\qquad E(N_n) \le C_1^n\cdot 2^{n^2(\gamma-\beta)d}\sum_{\underline{k}_1}\cdots\sum_{\underline{k}_n} \prod_{j=1}^{n-1} |\underline{k}_{j+1}-\underline{k}_j|^{-\gamma d/2}$$

$$\le C_2^n\cdot 2^{n^2(\gamma-\beta)d}.$$

On the event A_n,

$$(5.9)\qquad N_n \ge n\cdot\left[\frac{[2^{n\varepsilon d}]+1}{n}\right] \ge n\cdot\left[\frac{2^{n\varepsilon d}}{n}\right]^n.$$

Therefore,

$$P(A_n) \le C_2^n \cdot n^{n-1} \cdot 2^{n^2(\gamma-\beta-\varepsilon)d} \tag{5.10}$$

Since $\gamma-\beta < \varepsilon$, $\sum P(A_n) < \infty$. This completes the proof. ■

APPENDIX

We shall need some facts from the spectral theory of stationary random fields. Our main references are Gihman and Skorohod [15], Chapter IV, and Rozanov [30].

Let $X = \{(X_1(t)),\ldots,X_d(t)) \in R^d: t \in R^N\}$ be a measurable, d-dimensional, stationary Gaussian random field with mean zero and continuous covariance matrix

$$R_X(t) = (EX_k(s+t)X_\ell(s)) = \int_{R^N} e^{i\langle t,\lambda\rangle}\Delta(d\lambda),$$

where $\Delta(d\lambda) = (\Delta_{k\ell}(d\lambda))$ is a finite, positive definite $d\times d$ matrix valued measure. X has a stochastic integral representation

$$X(t) = \int_{R^N} e^{i\langle t,\lambda\rangle}W(d\lambda),$$

where $W(d\lambda)$ is a mean zero C^d-valued Gaussian random measure with

$$EW_k(A)\overline{W}_\ell(B) = \Delta_{k\ell}(A\cap B).$$

Let $\mathcal{H}_X$ denote the $L^2(P)$-closed complex linear span of $\{X_j(t): t \in R^N, 1 \le j \le d\}$ and let $L^2(\Delta)$ denote the space of complex d-dimensional measurable functions $f(\lambda) = (f_1(\lambda),\ldots,f_d(\lambda))$ on R^N with norm

$$\|f\|_\Delta^2 = \int\langle f(\lambda),\ \Delta(d\lambda)f(\lambda)\rangle.$$

The correspondence $f \longleftrightarrow \int\langle f(\lambda), W(d\lambda)\rangle$ is an isometry between $L^2(\Delta)$ and $\mathcal{H}_X$. Write

$$\mathcal{F}(X) = \sigma\{X(t)\colon t \in R^N\}.$$

For fixed $t \in R^N$, the linear shift operator

$$T(t)\colon L^2(\mathcal{F}(X)) \to L^2(\mathcal{F}(X))$$

is defined by

$$T(t)[f(X(t_1),\ldots,X(t_n))] = f(X(t_1+t),\ldots,X(t_n+t)).$$

For fixed t_0, $T(t_0)$ is said to be ergodic if the σ-field of invariant events

$$\mathcal{F}(X,t_0) = \{A \in \mathcal{F}(X)\colon T(t_0)1_A = 1_A \text{ a.s.}\}$$

is trivial, i.e., if every $A \in \mathcal{F}(X,t_0)$ has either probability zero or one.

The following result is well known in the case $N = 1$. (See [30].)

PROPOSITION A.1. $T(t_0)$ is ergodic if and only if for each $a \in R$

$$\Delta\{\lambda \in R^N\colon \langle t_0,\lambda\rangle = a\} = 0. \tag{A.1}$$

PROOF. Assume that for some fixed $a \in R$, (a.1) fails. Then for some $k \in \{1,\ldots,d\}$

$$Y = W_k\{\lambda \in R^N: \langle t_0,\lambda\rangle = a\}$$

is a nonzero Gaussian variable in $\mathcal{H}_x$.

$$T(t_0)Y = e^{ia}Y.$$

It follows that

$$T(t_0)|Y|^2 = (T(t_0)Y)(T(t_0)\overline{Y}) = (T(t_0)Y)(\overline{T(t_0)Y}) = |Y|^2.$$

Since $|Y|^2$ is not constant we see that $T(t_0)$ is not ergodic.

Conversely, suppose that (A.1) holds for all a. There exists a finite or countable collection of variables $\{Y_n\} \subseteq \mathcal{H}_x$ such that the sequences $\{T(kt_0)Y_n: -\infty < k < \infty\}$ span $\mathcal{H}_x$. Let

$$Y_n = \sum_{j=1}^{d} \int g_{n,j}(\lambda)W_j(d\lambda).$$

It follows that the stationary Gaussian sequence $\{T(kt_0)Y_n: -\infty < k < \infty\}$ has correlation function

$$\rho_n(k) = \int e^{i\langle kt_0,\lambda\rangle} \sum_{j,\ell} g_{n,j}(\lambda)\overline{g_{n,\ell}}(\lambda)\Delta_{j\ell}(d\lambda).$$

It follows from (A.1) that the sequence has a continuous spectral measure on R. For each n, let

$$\mathcal{F}_n = \sigma\{T(kt_0)Y_n: -\infty < k < \infty\}.$$

Let $T_n(t_0)$ denote the restriction of $T(t_0)$ to $L^2(\mathcal{F}_n)$. Then

$$L^2(\mathcal{F}(X)) \cong \underset{n}{\otimes} L^2(\mathcal{F}_n)$$

$$T(t_0) \cong \underset{n}{\otimes} T_n(t_0).$$

Here $\otimes$ denotes the tensor product and $\cong$ unitary equivalence. Each $T_n(t_0)$ is ergodic. (See [30], Chapter IV, §6.) Hence $T(t_0)$ is ergodic, too. ∎

Now let $\Delta = \Delta_d + \Delta_c$ be the decomposition of Δ into discrete and continuous parts. Write

(A.2) $$X(t) = X_d(t)+X_c(t),$$

where X_d and X_c are independent stationary Gaussian random fields with spectral measures Δ_d and Δ_c. Write

$$\mathcal{F}(X_d) = \sigma\{X_d(t): t \in R^n\}$$

$$\mathcal{F}(X_c) = \sigma\{X_c(t): t \in R^n\}.$$

It follows from (A.2) that

$$\mathcal{F}(X) \subseteq \mathcal{F}(X_d) \vee \mathcal{F}(X_c).$$

But it is also the case that $\mathcal{F}(X_d) \subseteq \mathcal{F}(X)$ which implies $\mathcal{F}(X_c) \subseteq \mathcal{F}(X)$. Hence

(A.3) $$\mathcal{F}(X) = \mathcal{F}(X_d) \vee \mathcal{F}(X_c).$$

Let $T_d(t)$ and $T_c(t)$ denote the restrictions of the shift operator $T(t)$ to $L^2(\mathcal{F}(X_d))$ and $L^2(\mathcal{F}(X_c))$, respectively. Since X_d and X_c are independent we have

(A.4) $$L^2(\mathcal{F}(X)) \cong L^2(\mathcal{F}(X_d)) \otimes L^2(\mathcal{F}(X_c))$$

$$T(t) \cong T_d(t) \otimes T_c(t)$$

Now consider a vector $t_0 \in R^N$ for which

(A.5) $$\Delta_c\{\lambda \in R^N: \langle t_o,\lambda\rangle = a\} = 0$$

for every $a \in R$. It follows from Proposition A.1 that $T_c(t_0)$ is ergodic. Consider the σ-field of invariant events

$$\mathcal{F}(X_d,t_0) = \{A \in \mathcal{F}(X_d): T(t_0)1_A = 1_A \text{ a.s.}\}.$$

It follows from (A.3) that $\mathcal{F}(X_d,t_0) \subseteq \mathcal{F}(X,t_0)$. Taking (A.4) into account we get

PROPOSITION A.2. If t_0 satisfies (A.5) for every $a \in R$, then modulo null sets

$$\mathcal{F}(X,t_0) = \mathcal{F}(X_d,t_0).$$

ACKNOWLEDGEMENT

This work was begun while the first author was visiting the Department of Mathematics of the University of Virginia. He wishes to thank the Department for its hospitality.

REFERENCES

1. R. J. Adler. Hausdorff dimension and Gaussian fields. Ann. Prob. 5 (1977), 145-151.

2. R. J. Adler. The uniform dimension of the level of sets of a Brownian sheet. Ann. Prob. 6 (1978), 509-515.

3. R. J. Adler. The Geometry of Random Fields. Wiley, New York, 1981.

4. S. M. Berman. Gaussian processes with stationary increments: local times and sample function properties. Ann. Math. Sat. 41 (1970), 1260-1272.

5. S. M. Berman. Gaussian sample functions. uniform dimension and Hölder conditions nowhere. Nagoya Math. J. 46 (1972), 63-86.

6. S. M. Berman. Local nondeterminism and local times of Gaussian processes. Indiana Univ. Math. J. 23 (1973), 69-94.

7. S. M. Berman. Gaussian processes with biconvex covariances. J. Multivariate Anal. 8 (1978), 30-44.

8. S. M. Berman. Local nondeterminism and local times of general stochastic processes. Ann. Inst. Fourier 19 (1983), 189-207.

9. J. M. Cuzick. Some local properties of Gaussian vector fields. Ann. Prob. 6 (1978), 984-994.

10. J. M. Cuzick. Local nondeterminism and the zeros of Gaussian processes. Ann. Prob. 6 (1978), 72-84.

11. J. M. Cuzick. Continuity of Gaussian local times. Ann. Prob. 10 (1982), 818-823.

12. J. M. Cuzick and J. P. DuPreez. Joint continuity of Gaussian local times. Ann. Prob. 10 (1982), 810-817.

13. O. Frostman. Potential d'équilibre et capacité des ensembles. Medd. Lunds Univ. Math. Sem. 3 (1935), 3-115.

14. D. Geman and J. Horowitz. Occupation densities. Ann. Prob. 8 (1980), 1-67.

15. I. I. Gihman and A. V. Skorohod. The Theory of Stochastic Processes I, Springer, New York, 1980.

16. J. Hawkes. Local properties of some Gaussian processes. Z. Wahrscheinlichkeitstheorie verw. Gebiete 40 (1977), 309-315.

17. J. Hawkes and W. E. Pruitt. Uniform dimension results for processes with independent increments. Z. Wahrscheinlichkeitstheories 28 (1974), 277-288.

18. J. P. Kahane. Ensembles aléatoires et dimensions. Prepublications 33, Univ. Paris-Sud Dept. Math., Orsay, France.

19. J. P. Kahane. Some Random Series of Functions, 2nd Ed., Cambridge University Press, Great Britain, 1985.

20. R. Kaufman. Une propriété métrique du mouvement brownien. C. R. Acad. Sc. Paris Sér. A-B 268 (1969), 727-728.

21. R. Kaufman. Brownian motion and dimension of perfect sets. Can. J. Math. 22 (1970), 674-680.

22. R. Kaufman. Temps locaux et dimensions. C. R. Acad. Sc. Paris Sér. I 300 (1985), 281-282.

23. U. Krengel. Ergodic Theorems. de Gruyter Studies in Mathematics 6, Berlin, 1985.

24. M. B. Marcus. Capacity of level sets of certain stochastic processes. Z. Wahrscheinlichkeitstheorie 34 (1976), 279-284.

25. E. Mourier. Elements aléatoires dans un espace de Banach. Ann. Inst. H. Poincaré Sect. B 13 (1953), 161-244.

26. J. Nolan. Local nondeterminism and local times for stable processes, preprint (1986), Univ. North Florida, Dept. of Math.

27. S. Orey. Gaussian sample functions and the Hausdorff dimension of level crossings. Z. Wahrscheinlichkeitstheorie 15 (1970), 249-256.

28. L. D. Pitt. Local times for Gaussian vector fields. Indiana Univ. Math. J. 27 (1978), 309-330.

29. C. A. Rogers. Hausdorff Measures. Cambridge Univ. Press, Great Britain, 1970.

30. Yu. A. Rozanov. Stationary Random Processes. Indian Statistical Institute, Calcutta, India.

31. F. Testard. Processus Gaussiens: polarite, points multiples, geometrie (1986). Pub. de Laboratori de Stat. et Prob. (No. 01-86) Univ. P. Sabatier Toulouse.

32. S. J. Taylor. The α-dimensional measure of the graph and the set of zeros of a Brownian path., Proc. Cambridge Philos. Soc. 51, Part II (1955), 265-274.

Department of Mathematics
University of Illinois
Altgeld Hall
1409 W. Green Street
Urbana, IL 61801

Department of Mathematics
University of Virginia
Math/Astronomy Bldg.
Cabell Drive
Charlottesville, VA 22903

LAST EXIT TIME AND HARMONIC MEASURE FOR BROWNIAN MOTION IN R^d

Z. R. Pop-Stojanovic

The purpose of this paper is to establish a connection between the <u>last exit time distribution</u> and the so-called <u>harmonic measure</u> of the Brownian motion process in R^d . The main tool in handling this problem is the following Chung's formula [1] : let D be a bounded domain in R^d , $(X(t))$ the Brownian motion process in R^d , and τ_D the exit time from D, i.e., $\tau_D = \inf\{t>0;\ X(t)\in D^c\}$. For every bounded measurable function f on ∂D define H_D as:

(*) $$H_D f(x) = E^x[f(X(\tau_D))\ ;\ \tau_D < \infty]\ ,\ x\in \overline{D}\ .$$

Then, according to [1], a measure $H_D(x,.)$ defined on the boundary ∂D is called the <u>harmonic measure of x with respect to D</u> .

CONNECTION BETWEEN τ_D AND $H_D(x,.)$.

Let G be the Dirichlet Green function of the bounded domain D in R^d of the Brownian motion process $(X(t))$. Assume that there is a sequence of domains (D_n) , $D_n \subset \overline{D}$, $n = 1,2,\ldots$, which expands to D. Let μ_n be the equilibrium distribution for D_n , $n = 1,2,\ldots$. Let x be a regular point of ∂D. Then $G(x,y) \to 0$ as $y \to x$. Since

$$\int G(x,y)\mu_n(dy) \to 1 \text{ as } n \to \infty\ ,$$

the total mass of μ_n must tend to infinity as $n \to \infty$. Intuitively one should expect that the sequence of measures $(G(x,y)\mu_n(dy))_n$ should converge to the harmonic measure at x. Indeed, we shall show that this is the case. The main tool for showing this is the Chung's formula (*).

Let (D_n) be a sequence of domains increasing to D. Let μ_n

denote the equilibrium measure of D_n , for n = 1,2,... , i.e.,

$$P_{D_n} 1 = \int G(x,y)\mu_n (dy) , n = 1,2,... .$$

We have the following

<u>Theorem</u> . Let x be a regular point of ∂D. Then the sequence of measures $(G(x,y)\mu_n (dy))_n$ converges weakly to the harmonic measure at x.

<u>Proof</u> . For each function f continuous on $\overline{D}$,

(1) $$u_n(x) = \int f(y)G(x,y)\mu_n(dy) , n = 1,2,...,$$

is harmonic in D_n and bounded by the the supremum norm of f . By Harnack inequality and a diagonal procedure one can find a subsequence of (1) which converges uniformly on compact subsets of D. Using a dense subset of the unit ball of $C(\overline{D})$ one can conclude that there is a subsequence such that along this subsequence, (1) converges uniformly on compacts of D, for all f in the unit ball of $C(\overline{D})$. Observe that the limit is a harmonic function which is bounded. We are showing that its boundary values (at regular points) are f. Let $y \in \partial D$ be a regular point. It is sufficient to show that if x is close to y and n large, then $u_n(x)$ is close to f(y). Let γ_n be the last exit from D_n before exit from D.By Chung's formula (*),

(2) $$u_n(x) = E^x [f(X(\gamma_n) ; \gamma_n < \infty] .$$

Since y is a regular point the standard classical argument implies that if x is close to y , the exit time is small with large P^x-probability. For any n such that D_n contains x, γ_n is small with large probability. Then, $X(\gamma_n)$ is close to y and $u_n(x)$ close to f(y). Thus the limit

$$u(x) = \lim_n u_n(x)$$

is harmonic in D and assumes boundary values f. By uniqueness u must be the stochastic solution. Also, if μ_x is the weak limit (along the mentioned subsequence) of $(G(x,y)\mu_n (dy))_n$ one has:

$$u(x) = \int f(y)\mu_x (dy) .$$

Since the measure μ_x is concentrated on the boundary od D, μ_x is the harmonic measure. See, M. Rao [2]. This completes the proof.

CONNECTION WITH THE MARTIN KERNEL

Let us denote by χ the measure on ∂D given by

$$\chi\,(dy) = \int \mu_x(dy)dx$$

where μ_x is the harmonic measure at a regular boundary point x. For each $x \in \partial D$, μ_x is absolutely continuous relative to χ. Denote a density by F(x,y). By a result due to Doob, we may assume that F is (x,y)-measurable. Now if h is an χ-integrable function it is also μ_x-integrable for at least one x and hence

$$\int h(y)\,\mu_x(dy)$$

is harmonic. Density F(x,y) is essentially bounded as x varies in a compact set in D and $y \in \partial D$. If q is a bounded measurable function on ∂D

$$\int F(x,y)q(y)\,\chi(dy) = \int q(y)\,\mu_x(dy)$$

is harmonic in D. For any relatively compact open set A and the hitting time T of A^c, one has

$$\int P_T\ F(x,y)q(y)\,\chi(dy) = \int F(x,y)q(y)\,\chi(dy)$$

which shows that $P_T F(x,y)$ is also a density, and for each y, $P_T F(.,y)$ is harmonic in A. If $B \supset A$ and S relates to B as T relates to A,

$$P_T F(x,y) = P_S F(x,y)$$

for each $x \in A$, and for almost all y. By Fubini theorem, this holds for almost all y and for almost all $x \in A$. Since both functions are harmonic in A, we see that for almost all y,

$$P_S F(x,y) = P_T F(x,y) \text{ for all } x \in A.$$

Thus, one can choose F such that for every y, F(.,y) is harmonic in D, F(x,.) is a density of μ_x and $\mu_x(1) = 1$. Hence,

$$\int F(x,y)\,\chi(dy)dx = |D|\quad,$$

where $|D|$ denotes the volume of D. The last relation shows that for χ-almost all y the harmonic function F(.,y) is integrable in D.

Since $(G(x,y)\mu_n(dy))_n$ converges weakly to $\mu_x(dy)$, $(\sigma(y)\mu_n(dy))_n$ converges weakly to $\chi(dy)$ where

$$\sigma(y) = \int G(x,y)\, dx \, .$$

Write

$$G(x,y)\mu_n(dy) = (G(x,y)/\sigma(y))\sigma(y)\mu_n(dy).$$

As y tends to a Martin point z , $G(x,y)/\sigma(y)$ tends to the Martin kernel $K(x,z)$, showing the connection between the Martin kernel and the density F described earlier.

ACKNOWLEDGMENT. The author wishes to express his gratitude to Murali Rao for his valuable comments concerning this paper.

REFERENCES

[1] K.L. Chung, Lectures from Markov Processes to Brownian Motion , Grundlehren der mathematischen Wissenschaften 249, Springer-Verlag, New York - Heidelberg - Berlin, (1982).

[2] Murali Rao, On Harmonic Measure for Brownian Motion, to appear.

DEPARTMENT OF MATHEMATICS, UNIVERSITY OF FLORIDA, GAINESVILLE, FLORIDA, 32611.

SOME REMARKS ON CAPACITIES

by J. Steffens

0. Introduction

In the paper "Capacity theory without duality" [6] a purely probabilistic notion of capacity relative to a general right process and an excessive measure was introduced by means of the corresponding stationary process.

To recall the definitions let us assume the set-up of [6] set down in section 2 there. That is, $\mathbf{X} = (X_t)_{t \geq 0}$ is a right process on $(\mathbf{E},\boldsymbol{\mathcal{E}})$ with semigroup $(\mathbf{P_t})_{t \geq 0}$, and $\mathbf{m}$ is an excessive measure. $\mathbf{E_b^a}$ (= E with a birth point $\mathbf{a}$ and a death point $\mathbf{b}$ adjoined) is the state space of the corresponding process $\mathbf{Y} = (Y_t)_{t \in \mathbb{R}}$, which has random birth and death, which is stationary in time, and which is Markov with semigroup (P_t) with respect to the Kuznetzov measure $\mathbf{Q_m}$ (associated with m). For a set B in $\mathcal{E}$ we let τ_B (resp. λ_B) denote the **first hitting** (resp. **last exit**) **time** of B with respect to Y . Then we defined $\mathbf{C(B)} := Q_m(0 < \lambda_B \leq 1)$ and $\mathbf{\hat{C}(B)} := Q_m(0 < \tau_B \leq 1)$, and we called these the **capacity** (resp. **cocapacity**) of B in case B is **transient** (resp. **cotransient**), which means $Q_m(\lambda_B = \infty) = 0$ (resp. $Q_m(\tau_B = -\infty) = 0$).

These definitions were justified by results obtained when specializing to the situation of (weak) duality (see [4] or [5]), and moreover they were shown to be more general. The use of the associated stationary process replaces the 'lack of duality' in this general set-up. One difference from the situation under duality is that in general there do not exist (co-)capacitary measures. Instead, there are the measures η_B on E_b given by $Q_m(Y_{\lambda_B} \in \cdot\ ; 0 < \lambda_B \leq 1)$ and $\hat{\eta}_B$ on E^a given by $Q_m(Y_{\tau_B} \in \cdot\ ; 0 < \tau_B \leq 1)$ the total masses of which are $C(B)$ and $\hat{C}(B)$. The main result that started the investigations of [6] is that for both transient and cotransient B the capacity equals the cocapacity.

Besides, the above definitions suggest the following intuitive description: the cocapacity of a set B is the amount of mass first entering B during a unit time interval under a stationary flow, and the capacity the amount last exiting B , accordingly.

In [6] expressions for C and $\hat{C}$ in terms of the original data — i.e. the right process X , its semigroup (P_t) , and the excessive measure m — were discussed only very briefly. The main emphasis lay in investigations within the set-up of Q_m -theory.

This note is to give some more relations of (co-)capacities with the original data and to sum up several facts that are consequences of results in [6], but were not explicitly stated there. As far as possible the results will be formulated for the measures η_B and $\hat{\eta}_B$ rather than just for their total masses $C(B)$ and $\hat{C}(B)$. Along these lines we shall obtain a new (shorter) proof of "Spitzer's formula" as stated in (7.1) of [6].

Thanks to a comment of C. Dellacherie on our paper, our attention was led to the analytic notion of the energy functional and its relations with capacities. Those will be just briefly sketched here, a further discussion will appear in [7].

This note is divided into three parts, the first one concerns the cocapacity, the second one the capacity, and the last one contains some final remarks. Before getting started let us fix some more notation:

The **birth** and **death times** of Y are denoted by α and β, respectively; T_B and L_B denote the **hitting** and **last exit times** of B with respect to X, P_B denotes the **hitting kernel**, and $\Phi_B := P_B 1 = P^{\cdot}(T_B < \infty) = P^{\cdot}(L_B > 0)$. By $(U_q)_{q>0}$ we denote the **resolvent** of X, and by U its **potential kernel**. The mappings γ_t are defined in (2.1) of [6], they relate the path space of Y to that of X.

$\mathcal{E}^a$, $\mathcal{E}_b$, $\mathcal{E}^a_b$ denote the σ-algebras associated with E^a, E^b, E^a_b. As usual, functions f on E are assumed to be extended to E^a_b by $f(a) := f(b) := 0$. $\underline{\mathcal{E}}$, $p\underline{\mathcal{E}}$, $b\underline{\mathcal{E}}$, $pb\underline{\mathcal{E}}$ stand for the **$\mathcal{E}$-measurable** (resp. positive, bounded, positive and bounded $\mathcal{E}$-measurable) **functions** on E, accordingly for other σ-algebras.

If μ is an excessive measure, then μ_i, μ_p, μ_d, and μ_c denote its **invariant**, **purely excessive**, **dissipative**, and **conservative parts**, respectively, as defined in [3]; furthermore, following [3] we use **Exc**, **Inv**, **Pur**, **Dis**, and **Pot** to denote the classes of excessive, invariant, purely excessive, dissipative, and potential measures on $(E,\mathcal{E})$ with respect to (P_t).

1. The cocapacity

Throughout this section let $B \in \mathcal{E}$ be a set in the state space, and let m be an excessive measure. First recall the definition of Hunt's **balayage** operation (see [8]) as extended by Fitzsimmons and Maisonneuve in [3] to general excessive measures:

$$\mathbf{R_B m(f)} := Q_m(f\circ Y_t \ ; \ \tau_B < t) \qquad \text{for } f \in p\underline{\mathcal{E}} ,$$

which is independent of $t \in \mathbb{R}$ (see [6]-(2.10)). More generally than what was explicitly stated in [6] — where the interest was mostly in cotransient sets — the following is true: The purely excessive part of $R_B m$ is given by $\mathbf{(R_B m)_p(f)} := Q_m(f\circ Y_t \ ; \ -\infty < \tau_B < t)$, and the corresponding **entrance law** is $(\rho^B_t)_{t>0}$ with $\rho^B_t\mathbf{(f)} := Q_m(f\circ Y_{t+\tau_B} ; \ 0<\tau_B \leq 1)$

(see (4.7) and (2.11) of [6]).

In (4.10) and (4.11) of [6] the following relations were proved:

(1.1) $\hat{C}(B) = \uparrow\lim_{t\downarrow o} \rho_t^B(1)$;

(1.2) $\hat{C}(B) = \uparrow\lim_{n} \mu_n P_B 1$, if B is cotransient and $m_d = \uparrow\lim_{n} \mu_n U$.

For (1.1) cotransience of B is in fact not needed.

Recall the definition of the **energy functional** L , first introduced and discussed by Meyer in [9] (see also [2]) in the situation of transient processes where Exc = Dis . For arbitrary $m \in$ Exc and excessive functions f , L can be defined (as before) by

(1.3) $\mathbf{L(m,f)} := \sup\{ L(\pi,f) : \pi \in \text{Pot} , \pi \leq m \}$,

where for $\pi = \mu U \in$ Pot , $L(\pi,f) = L(\mu U,f) := \mu(f)$.

Then $L(m,f) = L(m_d,f)$; L is increasing in both variables, and $L(m,Uf) = m(f)$. Furthermore, if $m \in$ Pur and f excessive one obtains $L(m,f) = \uparrow\lim_{t\downarrow o} \frac{1}{t}(m - mP_t)(f)$; and if $m \in$ Exc and f purely excessive one obtains $L(m,f) = \uparrow\lim_{t\downarrow o} \frac{1}{t} m(f - P_t f)$.

<u>**(1.4) Remark:**</u> *Since* $(R_B m)_p = \uparrow\lim_{t\downarrow o} \rho_t^B U$ *one has for any excessive* f ,

$$\uparrow\lim_{t\downarrow o} \rho_t^B(f) = \uparrow\lim_{t\downarrow o} L(\rho_t^B U,f) = L((R_B m)_p,f) .$$

In particular, for $f \equiv 1$ *it follows from* (1.1) *that*

(1.5) $\hat{C}(B) = L((R_B m)_p,1)$,

or for the ***cocapacity*** *of* B *in case* B *is* ***cotransient*** *(using* [3]-5.8)

(1.6) $\hat{C}(B) = L(R_B m,1) = L((R_B m)_d,1) = L(R_B m_d,1)$.

The following listing contains (1.1) and (1.2) and immediate consequences thereof. We give another proof using the energy functional.

(1.7) Corollary:

(1.8) $\hat{C}(B) = \mu P_B 1 = P^{\mu}(T_B < \infty)$, *if* $m = \mu U \in \mathrm{Pot}$;

(1.9) $\hat{C}(B) = \uparrow\lim_{t\downarrow o} \nu_t P_B 1 = \uparrow\lim_{t\downarrow o} P^{\nu_t}(T_B < \infty)$, *if* $m = \int_0^\infty \nu_t dt \in \mathrm{Pur}$;

(1.10) $\hat{C}(B) = \uparrow\lim_{k} (R_B m)_p(f_k)$, *if* $m \in \mathrm{Dis}$ *and* $Uf_k \uparrow 1$ a.e. m ;

(1.11) $\hat{C}(B) = \uparrow\lim_{n} \mu_n P_B 1 = \uparrow\lim_{n} P^{\mu_n}(T_B < \infty)$, *if* $\mu_n U \uparrow m \in \mathrm{Dis}$ *and* B *is cotransient.*

Proof: Recall from [3]-(5.8) that if $\mu_n U \uparrow m$,then $\mu_n P_B U \uparrow R_B m$. Hence, if $m = \mu U$, then $R_B m = \mu P_B U \in \mathrm{Pur}$, and by (1.5)

$$\hat{C}(B) = L(\mu P_B U, 1) = \mu P_B 1 .$$

If $m = \int_0^\infty \nu_t dt \in \mathrm{Pur}$, then $m = \uparrow\lim_{t\downarrow o} \nu_t U$, and $R_B m \in \mathrm{Pur}$ as well; therefore again by (1.5)

$$\hat{C}(B) = L(R_B m, 1) = \uparrow\lim_{t\downarrow o} L(\nu_t P_B U, 1) = \uparrow\lim_{t\downarrow o} \nu_t P_B 1 .$$

(1.10) follows from (1.5) as well:

$$\hat{C}(B) = L((R_B m)_p, 1) = \uparrow\lim_{k} L((R_B m)_p, Uf_k) = \uparrow\lim_{k} (R_B m)_p(f_k) .$$

(1.11) follows from (1.6):

$$\hat{C}(B) = L(R_B m, 1) = \uparrow\lim_{n} L(\mu_n P_B U, 1) = \uparrow\lim_{n} \mu_n P_B 1 .$$ //

The results of (1.7) cannot in general be formulated for the measure η_B on E^a , but only for $\hat{\eta}_B(f)$ such that $f \in p\mathcal{E}^a$ and the function $t \to f \circ Y_t$ exhibits some regularity at τ_B on $\{ \tau_B = \alpha \}$ a.s. Q_m . Nevertheless, the restriction of $\hat{\eta}_B$ to E can be expressed in terms of the initial data. Before stating this let us first recollect some facts which are formulated in section 7 of [6]. Let $\nu := m - R_B m$ and $\nu^i := m_i - R_B m_i$ (m_i is the invariant part of m .) . For $f \in p\mathcal{E}$ we denote the function $t \to P^{\nu}(f \circ X_{T_B} ; 0 < T_B \leq t)$ for $t > 0$, by u_m (ν is defined in terms of m !) . Finally, let $(K_t)_{t>o}$ denote the semigroup of the process X killed at T_B .

(1.12) Remarks: 1.) ν *is a well defined σ-finite positive measure with* $\nu(f) = Q_m(f\circ Y_t\ ;\ \alpha < t < \tau_B)$ *and* $\nu^i(f) = Q_{m_i}(f\circ Y_t\ ;\ t < \tau_B)$. ν *is* (K_t)*-excessive, and* ν^i *is the* (K_t)*-invariant part of* ν *since* $\nu K_t f = Q_m(f\circ Y_0\ ;\ 0 < \tau_B\ ;\ \alpha < -t)$.

2.) *Furthermore,* u_m *is increasing and subadditive;* u_m *is additive if* m *is invariant. This follows from 1.) since then* $\alpha = -\infty$ a.s. Q_m *and* $P^\nu(f\circ X_{T_B}\ ;\ s < T_B \leq s+t) = P^{\nu K_s}(f\circ X_{T_B}\ ;\ 0 < T_B \leq t)$.

3.) *Hence* u_m *is finite if it is finite for some* $t > 0$; *and besides* u_m *is right continuous.*

4.) *If* m *is invariant, then from the additivity and right continuity of* u_m *one has* $u_m(t) = t\cdot u_m(1)$ *for all* $t > 0$.

5.) *It follows as a standard fact on subadditive functions (as in section 8 of* [5]) *that* $\lim_{t\downarrow 0} \frac{1}{t}u_m(t)$ *exists and equals the increasing limit* $\uparrow\lim_n 2^n\cdot u_m(2^{-n})$. *If in particular* m *is invariant, then* $\lim_{t\downarrow 0} \frac{1}{t}\cdot u_m(t) = u_m(1)$.

(1.13) Proposition: *Let* $f \in p\underline{\mathcal{E}}$. *Then*

$$\hat{n}_{B|E}(f) = Q_m(f\circ Y_{\tau_B}\ ;\ 0 < \tau_B \leq 1\ ;\ \alpha < \tau_B)$$
$$= \lim_{t\downarrow 0} \frac{1}{t} P^\nu(f\circ X_{T_B}\ ;\ 0 < T_B \leq t) = \lim_{t\downarrow 0} \frac{1}{t}\cdot u_m(t) \ ,$$

and this limit is an increasing one along the sequence $t_k := 2^{-k}$ $(k\in\mathbb{N})$.

Proof: As in the proof of (7.12) of [6] we have

$$Q_m(f\circ Y_{\tau_B}\ ;\ 0 < \tau_B \leq 1\ ;\ \alpha < \tau_B) = \uparrow\lim_n \sum_{k=0}^{2^n-1} Q_m(f\circ Y_{\tau_B}\ ;\ \alpha < \frac{k}{2^n} < \tau_B \leq \frac{k+1}{2^n})$$

$$= \uparrow\lim_n 2^n\cdot Q_m(f\circ Y_{\tau_B}\ ;\ \alpha < 0 < \tau_B \leq \frac{1}{2^n}) = \uparrow\lim_n 2^n\cdot Q_m[P^{Y_0}(f\circ X_{T_B};\ 0<T_B\leq \frac{1}{2^n});\ \alpha < 0 < \tau_B]$$

$$= \uparrow\lim_n 2^n\cdot P^\nu(f\circ X_{T_B}\ ;\ 0 < T_B \leq \frac{1}{2^n}) = \uparrow\lim_n 2^n\cdot u_m(2^{-n})\ ;$$

and the result follows according to (1.12.5) . //

(1.14) Remarks: 1.) *The measure* $\hat{n}_B$ *on* E^a *in general is given by*

$$\hat{n}_B(f) = \lim_{t\downarrow o} \frac{1}{t}\cdot u_m(t) + f(a)\cdot Q_m(0 < \tau_B \leq 1 \,;\, \alpha = \tau_B) \quad \textit{for} \quad f \in p\underline{\underline{\mathcal{E}}}^a .$$

2.) $\hat{n}_B$ *is carried by* E *if on* $\{\tau_B > -\infty\}$ $\alpha < \tau_B$ *holds* a.s. Q_m . *This is the case in particular if* m *is invariant or if* B *is strongly* m *-cotransient, i.e.* $Q_m(\tau_B = \alpha) = 0$.

(1.15) Corollary: *If* m *is invariant, and again* $\nu = m - R_B m$ *, then for* $f \in p\underline{\underline{\mathcal{E}}}$, $\hat{n}_B(f) = P^\nu(f\circ X_{T_B} \,;\, 0 < T_B \leq 1)$, *in particular,*

(1.16) $\quad \hat{C}(B) = P^\nu(0 < T_B \leq 1)$.

Proof: This is immediate from (1.14.2) and (1.12.4), besides it was part of (7.12) of [6]. //

Finally, we give another proof of the **Spitzer formula** as formulated in (7.1) of [6]. For this it is essential that ν^i is the (K_t)-invariant part not only of ν (as stated in (1.12.1)), but of m itself.

(1.17) Lemma: *Let as before* m_i *denote the invariant part of* m , $\nu := m - R_B m$ *, and* $\nu^i := m_i - R_B m_i$ *, then* ν^i *is the* (K_t)*-invariant part of* m *(for the semigroup* (K_t) *of the process* X *killed at* T_B*).*

Proof: Let $f \in p\underline{\underline{\mathcal{E}}}$ such that $m(f) < \infty$. Then for $t \geq 0$,

$$mK_t(f) = Q_m[P^{Y_s}(f\circ X_t \,;\, t < T_B) \,;\, \alpha < s] =$$

$$= Q_m(f\circ Y_{s+t};\ s+t<\tau_B;\ \alpha<s) + Q_m(f\circ Y_{s+t};\ t<T_B\circ\gamma_s;\ \tau_B<s;\ \alpha<s) =$$

$$= Q_m(f\circ Y_o;\ 0<\tau_B;\ \alpha<-t) + Q_m(f\circ Y_o;\ 0<-t+T_B\circ\gamma_{-t};\ \tau_B<-t) .$$

The sets $\{0 < -t+T_B\circ\gamma_{-t}\} \cap \{\tau_B < -t\}$ are decreasing as $t \to \infty$ with intersection $\bigcap_{r>-\infty} \{r+T_B\circ\gamma_r > 0\} \cap \{\tau_B = -\infty\}$, which is empty.

Therefore we conclude $\lim_{t\to\infty} mK_t f = Q_{m_i}(f\circ Y_o \,;\, 0 < \tau_B) = (m_i - R_B m_i)(f)$.

//

What is proved implicitly in (1.17), in addition to (1.12.1), is that $R_B m$ is (K_t)-'purely supermedian', i.e. $R_B m K_t \leq R_B m$ and $R_B m(K_t f) \to 0$ as $t \to \infty$ for functions f with $R_B m(f) < \infty$. This was shown as well in [6] to obtain (7.16) there.

(1.18) Theorem (Spitzer's formula): *Suppose that* $P^m(0 < T_B \leq t)$ *is finite for some* $t > 0$, *then if* $f \in b\underline{\mathcal{E}}$,

$$\textbf{(1.19)} \qquad \lim_{t\to\infty} \frac{1}{t} P^m(f\circ X_{T_B} ; 0 < T_B \leq t) = P^{\nu^i}(f\circ X_{T_B} ; 0 < T_B \leq 1) = \hat{n}_B^i(f)$$

where $\hat{n}_B^i$ *is the measure on* E *corresponding to the invariant part* m_i *of* m.

Proof: We first show that $\lim_{t\to\infty} P^m(f\circ X_{T_B} ; t < T_B \leq t+1) = \hat{n}_B^i(f)$, the result then follows by 'Cêsaro-convergence'.

$P^m(f\circ X_{T_B} ; t < T_B \leq t+1) = P^m(f\circ X_{T_B}\circ \Theta_t ; 0 < T_B\circ\Theta_t \leq 1 ; t < T_B) =$
$P^m[P^{X_t}(f\circ X_{T_B} ; 0 < T_B \leq 1) ; t < T_B] = P^{mK_t}(f\circ X_{T_B} ; 0 < T_B \leq 1)$.

Now, $P^{\cdot}(f\circ X_{T_B} ; 0 < T_B \leq 1)$ is m-integrable by assumption (since the function $t \to P^m(0 < T_B \leq t)$ is subadditive, see (1.12)). Hence the last expression of the above equalities converges to $P^{\nu^i}(f\circ X_{T_B};0<T_B\leq 1)$ as $t \to \infty$, because according to (1.17) ν^i is the (K_t)-invariant part of m. This together with (1.15) establishes the first claim of the proof. As to the second part, observe that (with $[t]$ denoting the largest integer below t)

$$\frac{1}{t} P^m(f\circ X_{T_B} ; 0 < T_B \leq t) = \frac{[t]}{t} \Big[\frac{1}{[t]} P^m(f\circ X_{T_B} ; 0 < T_B \leq [t]) + \frac{1}{[t]} P^m(f\circ X_{T_B} ; [t] < T_B \leq t)\Big] ,$$

and note that as $t \to \infty$ the second term of the sum tends to zero since

$$P^m(f\circ X_{T_B} ; [t] < T_B \leq t) \leq P^m(f\circ X_{T_B} ; 0 < T_B \leq 1) .$$

Therefore $\lim_{t\to\infty} \frac{1}{t} P^m(f\circ X_{T_B} ; 0 < T_B \leq t) = \lim_{t\to\infty} \frac{1}{[t]} P^m(f\circ X_{T_B}; 0<T_B\leq[t]) =$

$$= \lim_{t\to\infty} \frac{1}{[t]} \sum_{k=0}^{[t]-1} P^m(f\circ X_{T_B}\ ;\ k < T_B \leq k+1) = \lim_{n\to\infty} P^m(f\circ X_{T_B}\ ;\ n < T_B \leq n+1)$$
$$= \hat{n}_B^i(f)\ . \qquad //$$

This proof is in fact shorter than the one we gave in [6]; what is needed for the first equality in (1.19) is only (1.12) and (1.17), and for the second, (1.13) and (1.15). Besides, in [6] the expression of the limit in terms of the initial data was not formulated. (Note that because of (1.17) the second expression in (1.19) can, in fact, be described just in terms of the initial data and does not require the notion of the stationary process!)

One final remark on cocapacities in connection with the considerations under (1.4) seems to be interesting:

(1.20) Remark: *Let* $\tilde{P}_B$ *denote the kernel on* E *which is given by*

$$\tilde{P}_B(f) := P^{\bullet}(f\circ X_{T_B}\ ;\ 0 < T_B < \infty)\ .$$

Then for any $f \in pb\underline{\underline{\mathcal{E}}}$ $\tilde{P}_B f$ *is purely excessive with respect to* (K_t), *because* $K_t\tilde{P}_B f = P^{\bullet}(f\circ X_{T_B}\ ;\ t < T_B < \infty)$. *In particular,* $\tilde{P}_B 1$ *is the* (K_t)*-purely excessive part of* $P^{\bullet}(T_B > 0) =: \tilde{1}$, *since* $K_t\tilde{1} = P^{\bullet}(T_B > t)$ *tends to* $P^{\bullet}(T_B = \infty)$ *as* $t \to \infty$. *Therefore, with the energy functional* $\tilde{L}$ *corresponding to the semigroup* (K_t), *we obtain from* (1.13),

$$\textbf{(1.21)} \qquad \hat{n}_B|_E(f) = \lim_{t\downarrow 0} \frac{1}{t} P^{\nu}(f\circ X_{T_B}\ ;\ 0 < T_B \leq t) =$$
$$= \uparrow\lim_{t\downarrow 0} \frac{1}{t}\, \nu(\tilde{P}_B f - K_t\tilde{P}_B f) = \tilde{L}(\nu,\tilde{P}_B f) \qquad (\text{for } f \in pb\underline{\underline{\mathcal{E}}})\,.$$

Consequently, if $\hat{n}_B$ *is carried by* E *(e.g. if* m *is invariant),*

$$\textbf{(1.22)} \qquad \hat{n}_B(f) = \tilde{L}(\nu,\tilde{P}_B f)\ ,$$

$$\textbf{(1.23)} \qquad \hat{C}(B) = \hat{n}_B(1) = \tilde{L}(\nu,\tilde{P}_B 1) = \tilde{L}(\nu,(\tilde{1})^P)\ ,$$

which is (1.22) *applied to* $f \equiv 1$, *where* $(\tilde{1})^P$ *denotes the* (K_t)*-purely excessive part of* 1.

2. The capacity

Again, throughout this section let $B \in \mathcal{E}$ and $m \in \text{Exc}$. In (4.3) of [6] the following relations were proved:

(2.1) $$C(B) = \lim_{t \downarrow 0} \frac{1}{t} P^m(0 < L_B \leq t) \;;$$

(2.2) $$C(B) = \lim_{t \downarrow 0} m(g_t^B) = \lim_{t \downarrow 0} m_d(g_t^B) \;, \text{ where } g_t^B := \frac{1}{t}(\Phi_B - P_t \Phi_B) \;.$$

(2.3) Remark: *Since for any finite excessive function* f *the purely excessive part* $(P_B f)_p$ *of* $P_B f$ *satisfies*

$$(P_B f)_p = \uparrow\lim_{t \downarrow 0} U f_t^B \;, \textit{ where } f_t^B := \frac{1}{t}(P_B f - P_t P_B f) \;, \textit{ one has for } f \;,$$

$$\uparrow\lim_{t \downarrow 0} m(f_t^B) = \uparrow\lim_{t \downarrow 0} L(m, U f_t^B) = L(m, (P_B f)_p) \;.$$

In particular, for $f \equiv 1$ *it follows from* (2.2) *that*

(2.4) $$C(B) = L(m, (\Phi_B)_p) \;,$$

or for the ***capacity*** *of* B *in case* B *is* ***cotransient***

(2.5) $$C(B) = L(m, \Phi_B) = L(m, P_B 1) \;.$$

Here, of course, L *denotes the* ***energy functional*** *again which was defined in* (1.3).

The subsequent remarks lead to generalizations of (2.1) and (2.2) for the measure η_B on E_b. Let $\mathbf{P^B}$ denote the '**last exit kernel**' on E_b, i.e. for $f \in p\underline{\underline{\mathcal{E}}}_b$ let $\mathbf{P^B(f)} := P^{\bullet}(f \circ X_{L_B} \;;\; 0 < L_B < \infty)$; let furthermore, for $f \in p\underline{\underline{\mathcal{E}}}_b$, the function $t \to P^m(f \circ X_{L_B} \;;\; 0 < L_B \leq t)$ for $t > 0$ be denoted by $\mathbf{v_m}$. Finally, let $\mathbf{L_t}$ abbreviate the operator $\frac{1}{t}(\text{Id} - P_t)$.

(2.6) Remarks: 1.) mP^B *is a measure on* E *if* B *is strongly* m-*transient, i.e. if* $Q_m(\lambda_B = \beta) = 0$ *or* $P^m(L_B = \zeta) = 0$ *with the life time* ζ *of* X .

2.) *For any* $f \in p b \mathcal{E}_b$, $P^B f$ *is a purely excessive function on* E *since* $P_t P^B f = P^{\bullet}(f \circ X_{L_B} ; t < L_B < \infty)$. *In particular, for* $\overline{1} := 1_{E_b}$, $P^B \overline{1} = P^{\bullet}(0 < L_B < \infty)$ *is the purely excessive part of* Φ_B , *because* $P_t P_B 1 = P^{\bullet}(L_B > t)$ *which decreases to* $P^{\bullet}(L_B = \infty)$ *as* $t \to \infty$, *and* $\Phi_B = P^{\bullet}(L_B > 0) = P^{\bullet}(L_B = \infty) + P^B \overline{1}$.

3.) *The function* v_m *has the same properties as* u_m *stated in* (1.12), *i.e. it is increasing, subadditive, and additive if* m *is invariant. The latter follows because*

$$P^m(f \circ X_{L_B} ; s < L_B \leq s+t) = P^{mP_s}(f \circ X_{L_B} ; 0 < L_B \leq t) .$$

Consequently, the properties (1.12.3 - 5) *are likewise valid with* u_m *replaced by* v_m .

4.) $L_t P^B f = \frac{1}{t} P^{\bullet}(f \circ X_{L_B} ; 0 < L_B \leq t)$ *for* $t > 0$, *therefore*

$$m(L_t P^B f) = \frac{1}{t} \cdot v_m(t) .$$

5.) $v_m = v_{m_d}$, *(because as in* [6]-(4.3)) *according to* [1], *the conservative part* m_c *of* m *does not contribute to the integral* $m(L_t P^B f)$.

With these facts established the following expressions for the measure η_B on E_b can be derived:

(2.7) Proposition: *Let* $f \in p\mathcal{E}_b$. *Then*

$$\eta_B(f) = Q_m(f \circ Y_{\lambda_B} ; 0 < \lambda_B \leq 1) = \lim_{t \downarrow 0} \frac{1}{t} P^m(f \circ X_{L_B} ; 0 < L_B \leq t) =$$

$$= \lim_{t \downarrow 0} m_d(L_t P^B f) = \lim_{t \downarrow 0} \frac{1}{t} \cdot v_m(t) ,$$

and this limit is an increasing one along the sequence $t_k := 2^{-k}$ $(k \in \mathbb{N})$.

Proof: The second equality holds according to the argument in (12.11) of [5] (quoted as well in [6]-(4.3)), which is completely analogous to the one for $\hat{\eta}_{B|E}$ in the proof of (1.13):

$$\eta_B(f) = Q_m(f \circ Y_{\lambda_B} ; 0 < \lambda_B \leq 1) = \uparrow\lim_n \sum_{k=0}^{2^n-1} Q_m(f \circ Y_{\lambda_B} ; \alpha < \frac{k}{2^n} < \lambda_B \leq \frac{k+1}{2^n})$$

$$= \uparrow\lim_n 2^n \cdot Q_m(f \circ Y_{\lambda_B} ; \alpha < 0 < \lambda_B \leq \frac{1}{2^n}) =$$

$$= \uparrow\lim_n 2^n \cdot Q_m[P^{Y_o}(f\circ X_{L_B} ; 0 < L_B \leq \frac{1}{2^n}) ; \alpha < 0] =$$

$$= \uparrow\lim_n 2^n \cdot P^m(f\circ X_{L_B} ; 0 < L_B \leq \frac{1}{2^n}) = \uparrow\lim_n 2^n \cdot v_m(2^{-n}) = \lim_{t\downarrow o} \frac{1}{t} \cdot v_m(t) .$$

Note that according to (3.4) of [6] $\lambda_B = L_B \circ \gamma_o$ on $\{\alpha < 0 < \lambda_B\}$ holds. The other equalities are immediate from (2.6.4/5). //

(2.8) Corollary: *If* m *is invariant, then for* $f \in p\underline{\underline{\mathcal{E}_b}}$, $\eta_B(f) = P^m(f\circ X_{L_B} ; 0 < L_B \leq 1)$, *in particular,*

(2.9) $$C(B) = P^m(0 < L_B \leq 1) .$$

Proof: If $m \in \mathrm{Inv}$, then according to (2.6.3) v_m is additive and $v_m(t) = t \cdot v_m(1)$. Hence (2.7) implies $\eta_B(f) = v_m(1) = P^m(f\circ X_{L_B} ; 0 < L_B \leq 1)$. Specializing to $f \equiv 1_{E_b}$ establishes (2.9). //

The result (2.7) in fact generalizes the relations (2.1), (2.2), and (2.4) for C(B) as explained in the following remark.

(2.10) Remark: *Since according to* (2.6.2) $P^B f$ *is purely excessive for any* $f \in pb\underline{\underline{\mathcal{E}_b}}$ *we obtain from* (2.7) *with the energy functional* L ,

(2.11) $$\eta_B(f) = \lim_{t\downarrow o} \frac{1}{t} P^m(f\circ X_{L_B} ; 0 < L_B \leq t) = \lim_{t\downarrow o} \frac{1}{t} m(P^B f - P_t P^B f) = L(m, P^B f) \qquad (\text{for } f \in pb\underline{\underline{\mathcal{E}_b}}) ;$$

(2.12) $$C(B) = \eta_B(\overline{1}) = L(m, P^B \overline{1}) = L(m, (\Phi_B)_p) ,$$

which is (2.11) *applied to* $f \equiv \overline{1} := 1_{E_b}$, *where* $(\Phi_B)_p$ *denotes the purely excessive part of* Φ_B .

From (2.11) expressions for η_B in various special cases can immediately be derived. However, since here we have not developed the properties of the energy functional in our set-up (without transience hypotheses; see [7]), we shall prove the results directly using (2.7).

(2.13) Theorem: *If* $m = \mu U \in \mathrm{Pot}$, *then the measure* η_B *on* E_b *is given by* μP^B .

Proof: For $m = \mu U$ and $f \in p\mathcal{E}_b$ the following equations hold:

$$\frac{1}{t} P^m(f \circ X_{L_B} ; 0 < L_B \leq t) = \frac{1}{t} P^\mu\Big[\int_0^\infty f \circ X_{L_B} \circ \Theta_s \, 1_{[0<L_B \circ \Theta_s \leq t]} \, ds \Big] =$$

$$= \frac{1}{t} P^\mu\big[f \circ X_{L_B} (L_B - (L_B - t) \vee 0) ; 0 < L_B < \infty \big] =$$

$$= \frac{1}{t} P^\mu(f \circ X_{L_B} \cdot L_B ; L_B \leq t) + P^\mu(f \circ X_{L_B} ; t < L_B < \infty) ,$$

and as $t \downarrow 0$ this sum tends to $P^\mu(f \circ X_{L_B} ; 0 < L_B < \infty)$. In fact, if $P^\mu(f \circ X_{L_B} ; 0 < L_B \leq t) = \infty$ for any $t > 0$, then

$$\lim_{t \downarrow 0} \frac{1}{t} P^m(f \circ X_{L_B} ; 0 < L_B \leq t) \geq P^\mu(f \circ X_{L_B} ; 0 < L_B < \infty) = \infty ;$$

If, however, it is finite for some $t > 0$, then

$$\frac{1}{t} P^\mu(f \circ X_{L_B} \cdot L_B ; L_B \leq t) \leq P^\mu(f \circ X_{L_B} ; 0 < L_B \leq t) ,$$ and thus tends to zero as $t \downarrow 0$. //

(2.14) Corollary: a) *If* $m = \uparrow\lim_n \mu_n U \in \mathrm{Dis}$, *then* $\eta_B = \uparrow\lim_n \mu_n P^B$.

b) *If* $m = \int_0^\infty \nu_t \, dt \in \mathrm{Pur}$ *with an entrance law* $(\nu_t)_{t>0}$,

then $\eta_B = \uparrow\lim_{s \downarrow 0} \nu_s P^B$.

Proof: Let $m = \uparrow\lim_n \mu_n U$. Define $t_k := 2^{-k}$ $(k \in \mathbb{N})$, then if $f \in p\mathcal{E}_b$, the numbers $a_{k,n} := \frac{1}{t_k} P^{\mu_n U}(f \circ X_{L_B} ; 0 < L_B \leq t_k)$ are increasing in both n and k , therefore by (2.7) and (2.13),

$$\eta_B(f) = \uparrow\lim_k \frac{1}{t_k} v_m(t_k) = \uparrow\lim_{k,n} a_{k,n} = \uparrow\lim_n \mu_n P^B f .$$

This establishes a). The statement in b) follows from a) since for $m \in \mathrm{Pur}$ as given one has $m = \uparrow\lim_{t \downarrow 0} \nu_t U$. //

The expressions for η_B in (2.14) lead, in particular, to formulas for $C(B)$ that are the analogues of those for $\hat{C}(B)$ stated in

(1.7). They will be summarized in the subsequent corollary. Except for the third formula (which — as well as (1.10) — was added in order to show the complete symmetry), they are immediate consequences of (2.14).

(2.15) Corollary:

(2.16) $C(B) = P^{B}\overline{1} = P^{\mu}(0 < L_B < \infty)$, *if* $m = \mu U \subset \text{Pot}$;

(2.17) $C(B) = \uparrow\lim_{t\downarrow 0} \nu_t P^{B}\overline{1} = \uparrow\lim_{t\downarrow 0} P^{\nu_t}(0<L_B<\infty)$, *if* $m = \int_0^\infty \nu_t \, dt \in \text{Pur}$;

(2.18) $C(B) = \uparrow\lim_{k} R_B m(f_k)$, *if* $m \in \text{Dis}$ *and* $Uf_k \uparrow 1$ a.e. m , *and* B *is transient;*

(2.19) $C(B) = \uparrow\lim_{n} \mu_n P^{B}\overline{1} = \uparrow\lim_{n} P^{\mu_n}(0<L_B<\infty)$, *if* $m = \uparrow\lim_{n} \mu_n U \in \text{Dis}$.

Proof: (2.16), (2.17), and (2.19) follow from (2.14). (Of course, using the energy functional, they are obtained as well from (2.12) — in the same way as the corresponding relations in (1.7) are proved.)
As to (2.18), note that if $m \in \text{Dis}$ then there exists a sequence of potentials $\mu_n U$ increasing to m ; besides, B transient implies with (2.6.2) that $P^{B}\overline{1} = P_B 1$ a.s. m and hence a.s. μ_n (see the proof of (4.5) in [6]). Therefore (using (2.19)) one has

$$C(B) = \uparrow\lim_{n} \mu_n P^{B}\overline{1} = \uparrow\lim_{n,k} \mu_n P^{B} Uf_k = \uparrow\lim_{k} R_B m(f_k) .$$

//

Finally, we state a relationship analogous to Spitzer's formula for the last exits and the measure η_B .

(2.20) Theorem: *Suppose that* $P^m(0 < L_B \leq t)$ *is finite for some* $t > 0$, *then if* $f \in b\mathcal{E}_b$,

(2.21) $\lim_{t\to\infty} \frac{1}{t} P^m(f\circ X_{L_B} ; 0 < L_B \leq t) = P^{m\ell}(f\circ X_{L_B}; 0<L_B\leq 1) = \eta_B^i(f)$,

where η_B^i *is the measure on* E_b *corresponding to the invariant part* m_i *of* m .

Proof: The proof works analogously as the one for the Spitzer formula (1.19); it is even simpler since here the semigroup (K_t) of the killed process is not involved. Under the given assumptions we have

$$P^m(f\circ X_{L_B}\,;\ t < L_B \leq t+1) = P^{mP_t}(f\circ X_{L_B}\,;\ 0 < L_B \leq 1)\ ,$$

which tends to $P^{m_i}(f\, X_{L_B}\,;\ 0 < L_B \leq 1)$ as $t \to \infty$, since $mP_t \to m_i$ on functions integrable with respect to m . Furthermore, from (2.8),

$$P^{m_i}(f\circ X_{L_B}\,;\ 0 < L_B \leq 1) = \eta_B^i(f)\ .$$

The result then follows by taking the Césaro limit as in the proof of (1.18). //

3. Concluding remarks

(3.1) In the preceding sections have been given several analogous expressions for C and $\hat{C}$ in terms of X , m , (P_t) , and (K_t) . (1.1) corresponds to (2.2); this was 'slightly indicated' in (4.10) of [6], here, however, by means of the remarks (1.4) and (2.3) the analogy is obvious. The formulas (2.16) - (2.19) are the analogues of (1.8) - (1.11). The counterpart of (1.16) is (2.9); and clearly the Spitzer formula (1.19) has its counterpart in (2.21). These relations give proof of the symmetry in the description of C and $\hat{C}$ also with respect to the initial data listed above.

(3.2) The symmetry is nicely represented in the analytic characterization of C and $\hat{C}$ by means of the energy functional, (1.5) and (2.4):

(3.3) $$C(B) = L(m,(P_B 1)_p) \qquad \text{and} \qquad \hat{C}(B) = L((R_B m)_p, 1)$$

Assume for the moment m dissipative, and let m be the increasing limit of potentials $\mu_n U$, and let 1 be the increasing limit of potentials Uf_k . Then from (3.3), formulas (1.10), (1.11), (2.18), (2.19) become apparent, and how they are related to each other is immediately clear (once one knows that $R_B m$ is the increasing limit of the potentials $\mu_n P_B U$):

$$C(B) = \uparrow\lim \mu_n(P_B 1)_p \qquad \text{and} \qquad \hat{C}(B) = \uparrow\lim (R_B m)_p f_k \ ;$$

furthermore,

$$C(B) = \uparrow\lim R_B m(f_k) \qquad \text{and} \qquad \hat{C}(B) = \uparrow\lim \mu_n(P_B 1)$$

$$\text{provided } P_B 1 = (P_B 1)_p \qquad \text{provided } R_B m = (R_B m)_p$$

Moreover, for both transient and cotransient B , it follows easily from (3.3) and the properties of L that

$$C(B) = L(m, P_B 1) = L(R_B m, 1) = \hat{C}(B) \ .$$

And last not least, the usual properties of capacities (as proved for C and $\hat{C}$ in (4.5) and (4.15) of [6]) can be derived at once from the corresponding properties of the functional L .

These observations give rise to another approach to capacity theory associated with a resolvent, that (together with its relations with Q_m-theory) will be discussed in [7] in wider generality. In particular, it will be shown that (for dissipative m) L is the proper extension of the set functions C and $\hat{C}$ on both transient and cotransient sets to general Borel sets. (See remark (4.16) in [6].)

<u>**(3.4)**</u> In general, in formulas for η_B and $\hat{\eta}_B$, the symmtry in the discription is disturbed — an effect due to the direction of time and the one-sided Markov property. For instance, there is the cocapacitary entrance law (ρ_t^B) (discussed at length in section 5 of [6]), a corresponding 'capacitary exit law' does not exist. On the other hand, there is the kernel P^B on E_b for the description of η_B , an analogous kernel on E^a defined in terms of the initial data does not exist either. Problems occur at the birth point a when the birth time α is not $-\infty$, which means that the asymmetry is caused by terms related to the purely excessive part of m ; to see this compare e.g. the relations in (1.13) and (1.15) with those in (2.7) and (2.8). Only, if m

is invariant they are completely analogous, and then (1.16) corresponds to (2.9). The analogy of the Spitzer formulas (1.19) and (2.21) is also due to the fact that they mainly deal with the measures η_B^i and $\hat{\eta}_B^i$ associated with the invariant part of m.

(3.5) Under the assumption of m being invariant, moreover, the kernels P^B and $\tilde{P}_B$ correspond to each other in a dual way, or rather the data (X,m,L_B,P^B,L) and $(\tilde{X},\nu,T_B,\tilde{P}_B,\tilde{L})$, where $\tilde{X}$ denotes the process X killed at T_B (see the definitions in(1.20)). This becomes apparent e.g. from the expressions (1.21), respectively (1.22), and (2.11), or from (1.23) and (2.12):

$$\textbf{(3.6)} \qquad C(B) = L(m,P^B 1) \qquad \text{and} \qquad \hat{C}(B) = \tilde{L}(\nu,\tilde{P}_B 1)$$

Hence, on the other hand there ought to be a process $\tilde{\tilde{X}}$ and associated objects that correspond to (X,m,T_B,P_B,L) and with respect to that one has an expression for $C(B)$ which formally (as in (3.6)) corresponds to (1.5).

(3.7) As to the expressions for C and $\hat{C}$ in the case where m is purely excessive: the results (2.14.b), respectively (2.17), and (1.9) follow as well from the subsequent observation as was pointed out to me by R.K. Getoor.

For $m = \int_0^\infty \nu_t \, dt$ one has $Q_m = \int_{-\infty}^\infty \Theta_t(Q_\nu) \, dt$ with the Kuznetzov measure Q_ν associated with the entrance law $(\nu_t)_{t>0}$ (cf. [6]-(2.9)). Hence, for any intrinsic stopping time τ the following holds $(g \in p\underline{\underline{\mathcal{E}}}_b^a)$:

$$Q_m(g \circ Y_\tau \; ; \; 0 < \tau \leq 1) = \int_{-\infty}^\infty Q_\nu(g \circ Y_\tau \circ \Theta_t \; ; \; 0 < \tau \circ \Theta_t \leq 1) \, dt =$$

$$Q_\nu\Big(\int_{-\infty}^\infty 1_{]0,1]}(\tau - t) \, dt \cdot g \circ Y_\tau\Big) = Q_\nu(g \circ Y_\tau \; ; \; \tau \in \mathbb{R}) = Q_\nu(g \circ Y_\tau \; ; \; 0 \leq \tau < \infty) \; .$$

Consequently for $\tau = \lambda_B$ and $f \in p\underline{\underline{\mathcal{E}}}_b$ we conclude

$$Q_m(f \circ Y_{\lambda_B} \; ; \; 0 < \lambda_B \leq 1) = Q_\nu(f \circ Y_{\lambda_B} \; ; \; 0 \leq \lambda_B < \infty) = Q_\nu(f \circ Y_{\lambda_B} \; ; \; 0 < \lambda_B < \infty)$$

$$\overset{(*)}{=} \uparrow\lim_{t\downarrow o} Q_\nu(f\circ X_{L_B}\circ\gamma_t \ ; \ 0 < L_B\circ\gamma_t < \infty) = \uparrow\lim_{t\downarrow o} Q^\nu(P^{\gamma_t}(f\circ X_{L_B} \ ; \ 0 < L_B < \infty))$$
$$= \uparrow\lim_{t\downarrow o} \nu_t P^B f \quad ,$$

where the equality marked by (*) holds because $(f\circ X_{L_B} 1_{\{o<L_B<\infty\}})\circ\gamma_t$ increases to $f\circ Y_{\lambda_B} 1_{\{o<\lambda_B<\infty\}}$ a.s. Q_ν as $t\downarrow o$.

On the other hand, for $\tau = \tau_B$ we do not obtain such a general expression because $(f\circ X_{T_B} 1_{\{T_B<\infty\}})\circ\gamma_t$ increases to $f\circ Y_{\tau_B}\cdot 1_{\{o\leq\tau_B<\infty\}}$ only under regularity assumptions on f ,e.g. for $f \equiv 1_{E_\Delta}$. Then in fact it is true that the sets $\{T_B\circ\gamma_t < \infty\}$ increase to $\{0 \leq \tau_B < \infty\}$ a.s. Q_ν as $t\downarrow o$. Therefore we can conclude:

$$Q_m(0 < \tau_B \leq 1) = Q_\nu(0 \leq \tau_B < \infty) = \uparrow\lim_{t\downarrow o} Q_\nu(T_B\circ\gamma_t < \infty) =$$
$$= \uparrow\lim_{t\downarrow o} Q_\nu(P^{\gamma_t}(T_B < \infty)) = \uparrow\lim_{t\downarrow o} P^{\nu_t}(T_B < \infty)$$

(3.8) Another remark of R.K. Getoor concerns the Spitzer formula: the proof of (1.18) carries over to the situation with general multiplicative functionals in place of the killing according to T_B .

References:

[1] R.M. Blumenthal: A decomposition of excessive measures. To appear in Seminar on Stochastic Processes 1985.

[2] C. Dellacherie, P.A. Meyer: Probabilités et Potentiel. Vol. IV, chap. XII/XIII, preprint.

[3] P.J. Fitzsimmons, B. Maisonneuve: Excessive measures and Markov processes with random birth and death. Probab. Th. Rel. Fields 72, 319-336, (1986).

[4] R.K. Getoor: Capacity theory and weak duality. Seminar on Stochastic Processes 1983, 97-130. Birkhäuser, Boston 1984.

[5] R.K. Getoor, M.J. Sharpe: Naturality, standardness, and weak duality for Markov processes. Z. Wahrscheinlichkeitstheorie verw. Geb. 67, 1-62 (1984).

[6] R.K. Getoor, J. Steffens: Capacity theory without duality. Submitted to Z. Wahrscheinlichkeitstheorie verw. Geb., accepted by Probab. Th. Rel. Fields.

[7] R.K. Getoor, J. Steffens: The energy functional, balayage, and capacity. In preparation.

[8] G.A. Hunt: Markov processes and potentials III. Ill. J. Math. 2, 151-213 (1958).

[9] P.A. Meyer: Note sur l'interpretation des mesures d'êquilibre. Sêminaire de Probabilitê VII, 210-216 (1973).

Jutta Steffens
Institut für Statistik
und Dokumentation
Universität Düsseldorf
Universitätsstr. 1

D-4000 Düsseldorf 1

West-Germany

CORRECTION

F. B. Knight

In our paper [1], we obtained a representation of a general $x \in L_0^2(Z_t)$ as a sum of stochastic integrals with respect to independent Brownian and compensated Poisson processes (Theorem 2.4). The principal hypothesis was H3: all martingales are strict. At the end (Theorem 2.9) we asserted a converse to the effect that strictness is necessary for the representation. Unfortunately, the proof has a gap and the assertion is false. The following is a counterexample. Let $Z_t = \sigma(P_1(s), P_2(s), s \leq t)$, where P_1 and P_2 are independent Poisson processes. Since (P_1, P_2) is a 2-basis, the stochastic integral representation is well-known to hold in this case [3, Chapter III, T9], so the converse would imply that $P_1(t)-t$ is a strict martingale. Then for any optional $T<\infty$, it would follow that $P_1(T) \in Z_{T-}$ (strictness implies $Z_{T-} = Z_T$, [1,p.119]). However, let T be the minimum of the first jump times of P_1 and P_2. Then the set $\{P_1(T)-P_1(T-) = 1\}$ is in $\sigma\{Z_T\}$ but not in $\sigma\{Z_{T-}\}$. The converse fails. The same applies to the "Final Remark" of [2], which must be deleted.

1. F. B. Knight. On strict-sense forms of the Hida-Cramer representation. Seminar on Stochastic Processes 1984. Birkhauser, Boston (1986) pp. 109-137.

2. F. B. Knight. Poisson representation of strict regular step filtrations. In Seminaire de Probabilites XX, Lecture Notes in Mathematics, Springer-Verlag (1986).

3. P. Bremaud. Points Processes and Oueues, Springer-Verlag (1981).